The Foundations of Magnetic Recording

The Foundations of Magnetic Recording

John C. Mallinson

Center for Magnetic Recording Research
University of California, San Diego

ACADEMIC PRESS, INC.

Harcourt Brace Jovanovich, Publishers

San Diego New York Berkeley Boston
London Sydney Tokyo Toronto

ACADEMIC PRESS, INC.
1250 Sixth Avenue, San Diego, California 92101

United Kingdom Edition published by
ACADEMIC PRESS INC. (LONDON) LTD.
24–28 Oval Road, London NW1 7DX

Library of Congress Cataloging in Publication Data

Mallinson, John C.
The foundations of magnetic recording.

Includes index.
1. Magnetic recorders and recording. I. Title.
TK7881.6.M29 1987 621.38 87-1249
ISBN 0–12–466625–6 (alk. paper)

PRINTED IN THE UNITED STATES OF AMERICA

87 88 89 90 9 8 7 6 5 4 3 2 1

Contents

Chapter 3 Magnetic Recording Media

Chapter 4 Magnetic Recording Heads

Chapter 5 The Writing, or Recording, Process

Chapter 6 Reading, or Reproducing, Processes

Chapter 7 Noise Processes and Signal-to-Noise Ratios

Chapter 8 Audio and Instrumentation Recorders

Preface

For more than a decade, the field of magnetic recording has needed a comprehensive textbook that is suitable for use at the senior undergraduate or graduate level of study. With the recent establishment of academic centers in the United States, such as those at the University of California, San Diego, Carnegie-Mellon University, and the University of Santa Clara, where formal classroom instruction in magnetic recording science and technology is now being undertaken, the need has become more urgent. This book, which is based on the lecture notes of a class I have taught at least annually since 1972, is intended to satisfy this need.

No formal prerequisites are necessary for students of this book; however, the ideal student should have a firm grasp of undergraduate level physics (in particular electromagnetic theory), mathematics, chemistry, and electrical engineering and their interrelationships, because magnetic recording embraces all these disciplines. Above all, the reader of this book, no matter what may be his or her formal education, should be primarily interested in comprehending the physical nature of the subject. The principal emphasis in this book is to cover the most important basic topics in magnetic recording in a manner that is scientifically correct and readily understandable to the nonspecialist.

The book follows a logical sequence. In Chapters 1 and 2, the fundamental physics of and measurements in magnetism and magnetic materials are treated. In Chapters 3 and 4, the two major, unique components of a recorder, the media and the heads, are discussed. Chapter 5 deals with the write or recording process, which is, even today, poorly understood. Chapter 6 covers the read or reproduce process. The foundations of the theory of noise and signal-to-noise ratios are given in Chapter 7. Chapters 8, 9, and 10 cover audio-instrumentation, video, and digital recording systems, respectively. The student is led from basic physics to philosophical considerations of system design.

Considerable liberties have been taken with the mathematical notation. For example, no distinction is made between the positions of the reproducing head, x', and the recorded medium, x. Thus the Reciprocity Integral is given, in Chapter 6, as

$$\Delta = \int \mathbf{h} \cdot \mathbf{M} \, dV$$

and not in the correct, full notation as,

$$\phi(x') = \int \int \int \mathbf{h}(x + x') \cdot \mathbf{M}(x) dx \, dy \, dz$$

These simplifications are made because it is my opinion that the full notation impedes rather than aids the student's understanding of the physical essence of the problem. In the Reciprocity Integral, the integration is the important thing to understand and not the correct handling of "dummy" variables. Students will notice that vectors in the text are indicated by boldfaced type.

In general, this book does not give complete proofs of the many equations it contains; rather, the reader is merely guided through the several stages of an analysis. The final result is then given, and it is followed by an extended discussion, which emphasizes the physical nature and practical consequences of the findings.

The book is intended to be self-contained, and accordingly, references are not cited in the body of the work. A list of further reading material is, however, included for each chapter. These lists will direct the student to the original papers and to more detailed accounts of particular topics. On the other hand, a book of this size can make no pretense of covering all the current fields of research, investigation, or development. Readers will, however, be treated to a solid, basic understanding of the field of magnetic recording.

Dedication and Acknowledgment

This book is dedicated to the Ampex Corporation because it provided the author, from 1962 to 1984, with the environment in which most of the ideas contained herein could be understood and refined. In particular, thanks are due to Neal Bertram, Michael Felix, Robert Hunt, Charles Steele, and Roger Wood for the countless hours of lively debate and assistance they so willingly provided.

Chapter 1

B, H, and M Fields, Magnetism and Measurements

1.1 Introduction

To understand magnetic recording it is necessary to distinguish carefully the properties and differences between the fields **B**, **H**, and **M**. Each is a field which, at all points in three dimensional space, defines the magnitude and direction of a vector quantity. The field **B** is called the magnetic flux density, **H** the magnetic field, and **M** the magnetization. All have properties similar to those of other, perhaps more familiar, fields such as the water flow in a river, the airflow over a wing, or the earth's gravitation.

There exists some arbitrariness concerning the order in which **B**, **H**, and **M** are introduced. Here the magnetic field is defined first, because it can be related to simple classical experiments. Magnetism and magnetization are treated next as logical extensions of the concept of a magnetic field. Finally the magnetic flux density is discussed as the vector sum of the magnetic field and the magnetization.

1.2 The Magnetic Field, H, and Magnetic Moment, $\boldsymbol{\mu}$

Consider an infinitely long, straight conductor carrying electric current, I amperes, as shown in Figure 1.1. The current

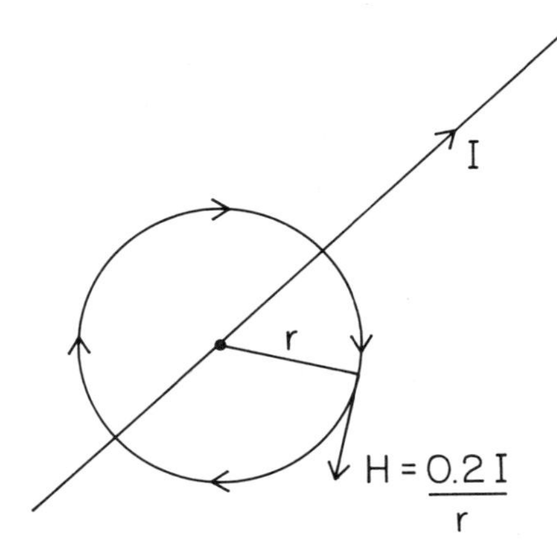

Fig. 1.1. The tangential field from a long filamentary conductor.

induces a magnetic field which circles the conductor. The magnetic field is everywhere tangential, that is, normal, to the radius. The direction of the magnetic field is that given by the right-hand rule: point the thumb of the right hand in the direction of the current flow and the magnetic field direction is given by the way the curved fingers point. The magnitude of the field is given by

$$H = \frac{0.2I}{r} \tag{1.1}$$

where H is the magnetic field in oersteds (Oe), I is the electric current in amperes (A), and r is the radial distance in centimeters (cm). All magnetic quantities in this book will be given in the centimeter, gram, second, electromagnetic (cgs–emu) system of units. Conversion factors to the Système Internationale of metric units (SI) are given in the Appendix.

Now suppose that the conductor is coiled to form the solenoid shown in Figure 1.2. The magnetic field inside a long solenoid is nearly uniform (parallel and of the same magnitude) and is

$$H = \frac{0.4\pi NI}{l} \tag{1.2}$$

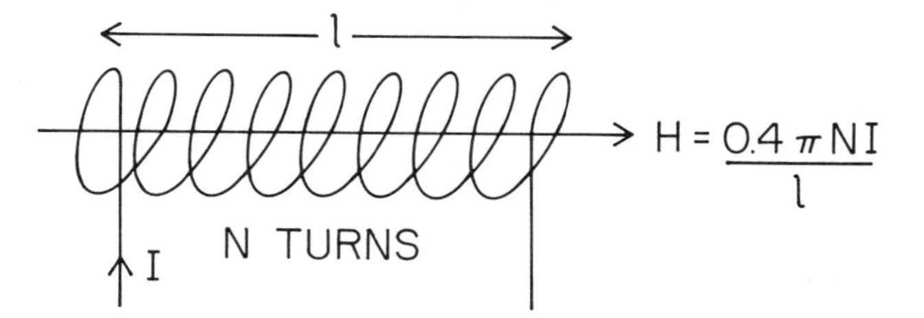

Fig. 1.2. The axial field from a long solenoid.

where H is the magnetic field in oersteds, N is the number of turns, I is the current in amperes, and l is the solenoid length in centimeters. Note that the field does not depend upon the cross-sectional area of a long solenoid.

At this point, it is convenient to define another vector quantity, the magnetic moment,

$$\boldsymbol{\mu} = \int_{inside\ coil} \mathbf{H}\, dv \tag{1.3}$$

where $\boldsymbol{\mu}$ is the magnetic moment in electromagnetic units (emu), and v is volume in cubic centimeters. For any solenoid, it may be shown that

$$\boldsymbol{\mu} = 0.4\pi NI\mathbf{A} \tag{1.4}$$

where $\mathbf{A}$ is the vector normal to the area of the solenoid cross section, whose magnitude is the solenoid cross-sectional area in square centimeters.

1.3 Magnetism and the Magnetization, M

Magnetization is a property which arises from the motion of electrons within atoms; consequently the magnetization of "free space," that is space free of material bodies, is by definition zero. In all atoms, electrons orbit a nucleus made up of protons and neutrons. The atomic number is the number of protons, which for an electrically neutral (not ionized) atom equals the number of electrons.

Within an atom, the electron has two separate motions. First, the electron orbits the nucleus, much as the earth orbits the sun, at a radius of a few angstroms (10^{-8} cm). Second, the electron spins on its own axis, much as does the earth. These two motions are, of course, governed always by quantum mechanical laws. For isolated atoms, these laws are known as Hund's rules. Any motion of an electron produces an electric current, and just as the motion of electrons around a solenoid produces a magnetic moment, the orbital motion gives rise to an orbital magnetic moment and the spinning motion causes an electron spin magnetic moment.

Generally, Hund's rules prescribe that the several electrons in an atom orbit in opposite directions so that the total orbital moment is small. In the solid state, the quantum mechanical interactions with neighboring atoms "quench," that is, reduce further, the orbital moment. For the materials used in magnetic recording (the first transition group of elements), the contribution of the orbital magnetic moments to the magnetization is virtually negligible.

The magnetic moment of a spinning electron is called the Bohr magneton and is of magnitude

$$\mu_B = \frac{e\hbar}{2m} = 0.93 \cdot 10^{-20} \text{ emu} \tag{1.5}$$

where e is the electron charge in electromagnetic units ($1.6.10^{-20}$), $\hbar$ is Plank's constant divided by 2π (1.05×10^{-27}), and m is the electron mass in grams (9×10^{-28}).

The spinning electron has a quantum spin number, $s = 1/2$, and can be oriented (in a weak magnetic reference field) in only two, $(2s + 1)$, directions. For atoms in free space, Hund's rules normally ensure that the electrons' spin directions alternate so that the total electron spin moment is no more than one Bohr magneton. Moreover, this small moment is reduced in the solid state by next-neighbor interactions. Fortunately, however, nature has provided irregularities in the electron spin ordering, and even better, has provided the transition elements. In transition elements, outer electron orbits, or shells, begin to be occupied before the inner ones are completely filled. The benefit of this is

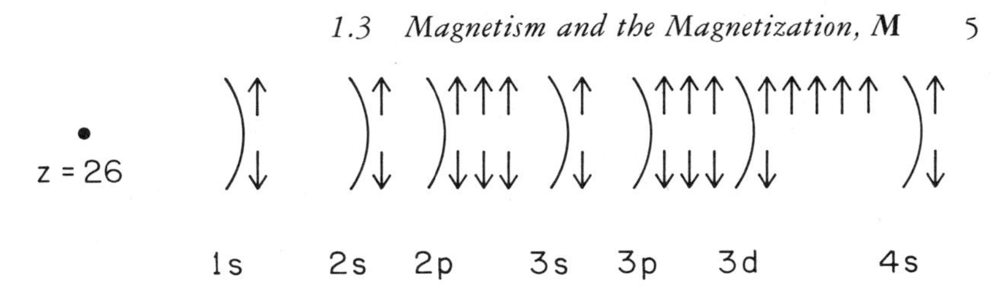

Fig. 1.3. The electron spin orientations of an iron atom in free space.

that the inner, partially filled shells can have large net electron spin moments and yet have the neighboring atom's interactions be partially screened off by their outer electrons.

Consider an atom of iron in free space, as depicted in Figure 1.3. The atomic number of iron is 26, and there are, accordingly, 26 electrons. The electron shells are shown numbered by two quantum numbers. The first, or principal, quantum number ($n = 1, 2, 3,$ and 4) relates to the electron's energy. The orbital quantum number (*s, p,* and *d*) defines the orbital shape. Note that in shells 1*s*, 2*s*, 2*p*, 3*s*, 3*p*, and 4*s* equal numbers of electron spins point up and down so that the total electron spin moment is zero; that is, the electron spins are "compensated." In shell 3*d,* however, an uncompensated spin moment of $4\mu_B$ exists because there are five spins up but only one down. Higher in the periodic table, the 3*d* shell eventually fills, resulting in ten electrons with five up and five down and a zero net electron spin moment. The first transition elements (Mn, Cr, Fe, Ni, and Co), however, have 3*d* shells unfilled and have uncompensated electron spin magnetic moments. Nearly all practical interest in magnetism centers on the first and second transition groups of elements with uncompensated spins.

When iron atoms condense to form a solid-state metallic crystal, the electronic distribution, called the density of states, changes. Whereas the isolated atom has 3*d*; 5+, 1−, 4*s*; 1+, 1−, in the solid state the distribution becomes 3*d*; 4.8+, 2.6−, 4*s*; 0.3+, 0.3−. Note that the total number of electrons remains eight, but that the uncompensated spin moment is lowered to $2.2\mu_B$. The screening of neighboring atom forces by the 4*s* electrons is imperfect. For all practical purposes, an iron atom in a magnetic material (pure iron,

iron alloy, or iron oxide) has $2.2\mu_B$ of magnetic or atomic moment. The electronic nature of the atomic moment will not be considered further here.

Now consider the magnetic behavior of iron atoms in an iron crystal. The crystal form, called the habit, is body-centered cubic with a cube-edge dimension of 2.86 Å (angstroms). First we ask what is the relationship of the atomic moment of one iron atom to those of its neighbors? It turns out that, at room temperature, all the iron atom magnetic moments point in nearly the same direction. A quantum mechanical force, called exchange, lowers the system's energy by aligning the uncompensated moments. At absolute zero, the ordering is perfect, while at higher temperatures, thermal energy causes increasing disorder.

At the Curie temperature (1100° C), thermal energy equals the exchange energy, and all long range order breaks down and the spin moments point randomly in all directions. In this disordered state, the material is said to be a paramagnet (i.e., almost a magnet). Below the Curie temperature, the parallel alignment is called ferromagnetism. Other orderings of the atomic moments are found in nature. In some materials, exchange forces cause each neighbor to be antiparallel; this is called anti-ferromagnetism, and of course, the material has zero total magnetic moment. In yet others, the number of moments in any one direction is not equal to those antiparallel; this intermediate case is called ferrimagnetism. Examples of ferromagnets are iron, cobalt, and chromium dioxide. The ferrimagnets include gamma-ferric oxide, ferrite magnetic head materials, and ferrite transformer materials. Example of anti-ferromagnets are manganese, alpha-ferric oxide, and cobalt oxide. The three important atomic moment orderings are depicted in Figure 1.4.

The second question to ask about iron in the solid state concerns the orientation of the net magnetic moment with respect to the crystal axes. It turns out that the ferromagnetic-ordered atomic moments align parallel to the body-centered cube edges. In iron the cube edges are the easy, or lowest, energy directions, with the body diagonals being the hard, or highest, energy directions of the magnetic moment.

Fig. 1.4. Three types of magnetic ordering.

A measure of this energy difference is the magnetocrystalline anisotropy constant, K. It is the energy required, in ergs per cubic centimeter, to rotate the magnetic moments from the easy to the hard direction. Several different symmetries of magnetocrystalline anisotropy are found in nature, but in magnetic recording most interest centers on cubic, as in iron and gamma-ferric oxide, and uniaxial, as in iron–nickel (e.g., permalloy) alloys and barium ferrite.

Now let our perspective expand to include a volume of iron which contains several million atoms. Just as previously our viewpoint moved from the electron spin level to the atomic level, now the focus is on millions of atoms. The magnetization is, by definition, the volume average of the atomic moments:

$$\mathbf{M} = \frac{1}{V} \sum_{i=1}^{N} \boldsymbol{\mu}_i \tag{1.6}$$

where V is the volume in cubic centimeters, $\boldsymbol{\mu}$ is the atomic magnetic moment in electromagnetic units, and N is the number of atomic moments in the volume V. The units of magnetization are, therefore, magnetic moment per unit volume or electromagnetic unit per cubic centimeter (emu/ cm^3).

In a large magnetic field, the magnetization at all parts of magnetic material is parallel; at lower fields, the magnetization may subdivide into domains. Domain behavior is treated in Chapter 2. Here it is noted only that within a domain the magnetization is everywhere parallel and uniform and has a value called the saturation magnetization, M_s.

The value of M_s depends on the temperature, being a maximum at absolute zero and vanishing at the Curie temperature.

The 0 K value of M_s for a body-centered cubic iron crystal can be calculated easily. Each iron atom has $2.2\mu_B$ of magnetic moment; there are, on average, two iron atoms per unit cell; and the cell edges measure 2.86 Å. It follows that

$$M_s(T = 0) = \frac{(2.2\mu_B)(2)}{(2.86 \times 10^{-8})^3} = 1700 \text{ emu/cm}^3 \quad (1.7)$$

At room temperature M_s is only slightly reduced, so for all practical purposes, pure iron has the following properties: $M_s = 1700$ emu/cm^3, $4\pi M_s = 21{,}000$ G (gauss), and $\sigma_s = 216$ emu/gm, where σ_s is called the specific saturation magnetization.

The values of $4\pi M_s$ for some other materials of interest in magnetic recording are Cobalt, 18,000 G; Nickel, 6,000 G; and Ferrites, 4–5000 G.

1.4 Demagnetizing Fields

In general, when a magnetic material becomes magnetized by the application of a magnetic field, it reacts by generating, within its volume, an opposing field that resists further increases in the magnetization. This opposing field is called the demagnetization field because it tends to reduce or decrease the magnetization.

In order to compute the demagnetization fields, first the magnetization at all points must be known. Then at all points within the sample one computes the magnetic pole density,

$$\rho = -\boldsymbol{\nabla} \cdot \mathbf{M} = -\left(\frac{dM_x}{dx} + \frac{dM_y}{dy} + \frac{dM_z}{dz}\right) \quad (1.8)$$

where ρ is the pole density (emu/cm^4), $\boldsymbol{\nabla}$ is the linear operator called divergence, and M_x, M_y, and M_z are the orthogonal components of the magnetization vector.

The convention for magnetic poles is that when the magnetization decreases the poles produced are north, or positive. It is an unfortunate historical accident that the earth's geographic north pole has south, or negative, magnetic polarity. The linear differential operator divergence is easy to visualize because it is merely the inflow or outflow of a vector field. Recall, for instance, that turning on the faucet in the bathtub makes the divergence at all points in the bath water flow field positive and pulling the plug makes the flow field divergence negative.

Magnetic poles are of extreme importance because they generate magnetic fields, **H**. There are only two sources of magnetic fields: real electric currents and magnetic poles. The adjective *real* is used to distinguish real currents flowing in wires, which may be measured with ammeters, from hypothetical currents flowing in atoms due to their orbiting and spinning electrons. These hypothetical currents are, confusedly enough, sometimes called Amperian currents.

The magnetic fields caused by magnetic poles may be computed using the Inverse Square law. The field points radially out from the positive, or north pole, and has the magnitude,

$$H = \frac{4\pi q}{r^2} \tag{1.9}$$

where H is the magnetic field in oersteds, q is the pole density times volume (emu/cm), and r is the radial distance in centimeters. As may be noticed, magnetic poles are analogous to electric poles.

Most field computations in magnetic recording are two dimensional. This is because one dimension, the track-width, is so large compared with the other two. Two dimensional magnetic fields have many simplifying properties, and one of the most important of these is shown as Figure 1.5. The magnetic field from a straight line of poles extending to $\pm\infty$ points radially out and has the magnitude,

$$H = \frac{0.2s}{r} \tag{1.10}$$

where H is the magnetic field in oersteds, s is the pole strength per unit length (emu/cm^2), and r is the radial distance in centimeters.

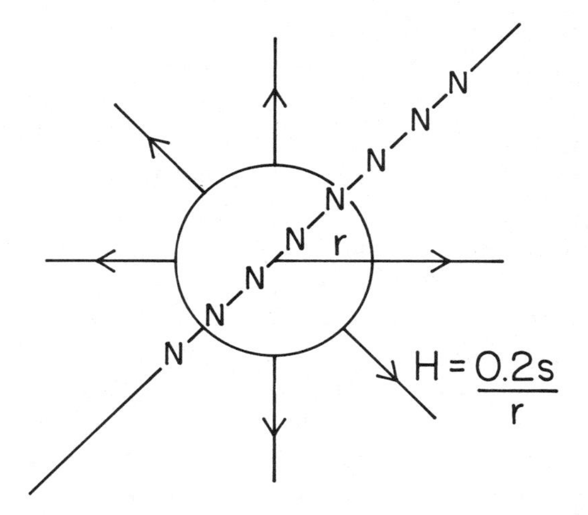

Fig. 1.5. The radial field from a long line of poles.

The crucial thing to notice is that Equations 1.1 and 1.10 have the same form. Apart from scaling factors, a change from electric current to magnetic poles causes a 90° rotation of the magnetic field at every point in a two-dimensional space. The magnetic fields from currents and poles are orthogonal.

In general, the fields generated by a magnetic body are very complicated and force the magnetization to be nonuniform. For one class of geometrical shapes, however, it is known that the demagnetizing field and magnetization can be uniform. When any ellipsoid of revolution is uniformly magnetized, the demagnetizing field is also uniform. The demagnetizing field can be written

$$\mathbf{H}_d = -\mathbf{N}\,\mathbf{M} \qquad (1.11)$$

where $\mathbf{H}_d$ is the vector demagnetizing field, $\mathbf{N}$ is the demagnetization tensor, and $\mathbf{M}$ is the vector magnetization. For ellipsoids of revolution, the demagnetization tensor is the same at all points within a given body. Note that since "tensors turn vectors" the demagnetizing field need not be exactly antiparallel to the magnetization.

Ellipsoids of revolution range from infinite flat plates

through oblate spheroids, to spheres, through prolate spheroids, through infinite cylinders. They are formed by rotating any ellipsoid about either its major or minor axis. The demagnetizing tensors for three cases are shown below.

tensor			flat plate			sphere			long cylinder		
xx	xy	xz	0	0	0	$\frac{4\pi}{3}$	0	0	2π	0	0
yx	yy	yz	0	0	0	0	$\frac{4\pi}{3}$	0	0	2π	0
zx	zy	zz	0	0	4π	0	0	$\frac{4\pi}{3}$	0	0	0

Thus the flat plate has no demagnetization within its x, y plane but suffers a 4π demagnetizing factor for magnetization components out of the plane. A sphere suffers a $4\pi/3$ factor in all directions. A long cylinder has no demagnetization along its axis, but suffers 2π in the x and y directions of its cross sections. Note that these tensors are all diagonal, because the axis of rotation coincides with the z direction, and that the diagonal terms always sum to 4π. This is because 4π steradians of solid angle fill three-dimensional space. In cgs–emu, 4π field lines emanate from a unit magnetic pole. In other systems of units, the 4π appears in other places, but its appearance cannot be suppressed.

Consider now an ellipsoidal sample of magnetic material within a long solenoid that produces an axial field, $\mathbf{H}_s$, as shown in Figure 1.6. Suppose that the solenoid field is strong enough to saturate the magnetization. The magnetization of the ellipsoid is uniform, divergence free, and there are no magnetic poles within the volume. Poles form, however, on the surfaces as shown by the letters N and S in the figure. These surface poles produce the demagnetizing field, $\mathbf{H}_d$, which is exactly antiparallel to both the magnetization and the solenoid field. If the ellipsoid axis had not been parallel to the solenoid axis these exact alignments would not have occurred. It is clear now that the total magnetic field within the sample is

$$\mathbf{H}_t = \mathbf{H}_s - \mathbf{H}_d \tag{1.12}$$

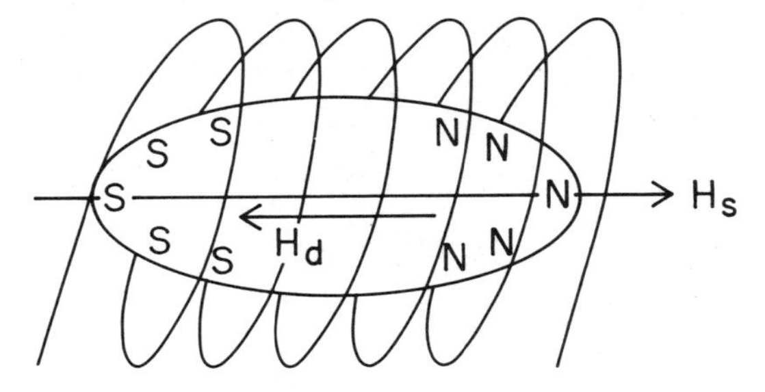

Fig. 1.6. An ellipsoid of revolution in a solenoid, showing the induced magnetic poles and the demagnetizing field.

The effect of the demagnetizing process is to change the field inside the sample in a manner which is exactly analogous to that of the negative feedback system shown in Figure 1.7.

1.5 The Flux Density, B

Having defined and discussed both the magnetic field **H** and the magnetization **M**, the flux density **B** can be defined as

$$\mathbf{B} = \mathbf{H} + 4\pi\mathbf{M} \tag{1.13}$$

where **B** is the flux density in gauss, **H** is the (total) field in oersteds, and **M** is the magnetization in electromagnetic units. In this equation, it is to be understood that **B, H,** and **M** are all field vector quantities and that the addition is performed vectorially (i.e., orthogonal component by component).

The **M** field, the **H** field, and the **B** field of a uniformly

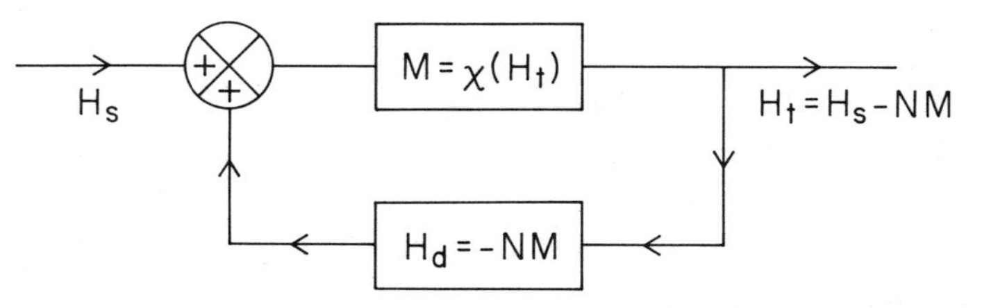

Fig. 1.7. Demagnetization represented as a negative feedback system.

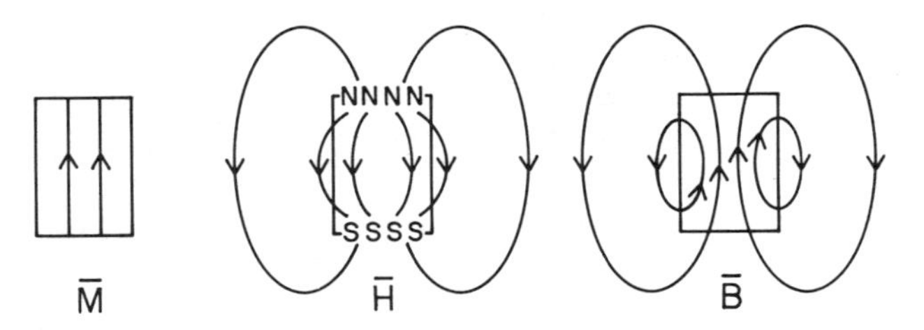

Fig. 1.8. Magnetization, magnetic field, and magnetic flux density fields of a bar magnet.

magnetized bar magnet are shown in Figure 1.8. In all the cases, the scheme adopted for showing, or plotting, the field in the plane of the paper is the same. The lines and arrows show, at all points, the direction of the particular field quantity. The spacing between the lines is inversely proportional to the field magnitude. The closer the lines, the higher the field strength. In this scheme, which is identical to that used to depict, for example, airflows, the "flow" or "flux" of the field quantity between all adjacent pairs of lines is equal. In dynamic fields, the lines are called streamlines; here the **B**, **H**, and **M** fields are static, but the same nomenclature persists. The streamlines are also called "lines of force," and between them, for the **B** field, flows the magnetic flux.

The **M** field pattern shows parallel lines within the uniformly magnetized magnet only. The magnetization outside the magnet is zero.

The **H** field drawing shows the magnetic poles caused by the changing magnetization on the ends of the magnet. Where **M** is decreasing, north, or positive, poles arise as at the top of the magnet. The convention is that **H** lines of force emanate from the north, or positive, poles. Note that within the body of the magnet, the magnetic field looks very similar to the electric field between two opposite-charged capacitor plates. Note also that the magnetic field within is not uniform but actually diverges or bows out; this is because the bar magnet's shape is not that of an ellipsoid of revolution. Note, further, that the magnetic field inside the magnet generally opposes the magnetization and is, appropriately enough, called the demagnetizing field. The magnetic field outside the magnet, which arises from the very same magnetic poles causing the demagnetizing field within, is often called the fringing field. Since both have the same origin, it is clear that larger external fringing fields imply higher internal demagnetizing fields.

The **B** field diagram is, of course, identical to that of the **H** field for all the points outside the magnet. This is because in free space, $\mathbf{B} = \mathbf{H}$, there being no magnetization. It follows that it is immaterial whether one speaks of the field in the gap of an electromagnet or a magnetic writing head as being **B** in gauss or **H** in oersteds; they are indistinguishable by experiment. Within the magnet, the vector addition of **H** and $4\pi\mathbf{M}$ produces the converging flow shown. Note, particularly, that unlike the **H** and **M** fields, the **B** field has no sources or sinks. This is expressed mathematically by writing $\nabla \cdot \mathbf{B} = 0$ and it is a consequence of the fact that isolated magnetic poles cannot exist; north and south poles exist in equal numbers.

1.6 Magnetic Measurements

There are three basic types of equipment used to characterize the magnetic properties of the materials used in magnetic recording. They are the 60-Hz M–H looper, the toroidal B–H looper, and the vibrating sample magnetometer (VSM), which measures M–H and other loops. The properties of these hysteresis loops are discussed in Chapter 2, here only the types of equipment used are discussed.

Figure 1.9 shows the layout of a 60-Hz looper. Within a long solenoid, which is driven by 60-Hz house current, are two nearly identical pick-up coils. The polarities of electrical connection of these two coils are opposite so that the voltages induced in them by the 60-Hz solenoid field cancel out. These voltages are given by Faraday's law,

$$E = -10^{-8} N \frac{d\phi}{dt} \tag{1.14}$$

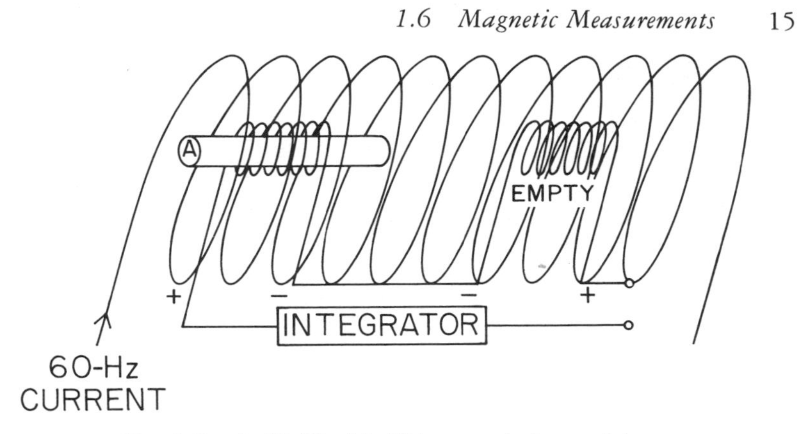

Fig. 1.9. A 60 Hz *M–H* looper (schematic).

where E is the voltage in volts, N is the number of turns in the coil, ϕ is the magnetic flux ($\mathrm{G - cm^2}$), and t is time in seconds. The magnetic flux ϕ is given by

$$\phi = \int \mathbf{B} \cdot d\mathbf{A} \tag{1.15}$$

and, in the empty pick-up coil, is equal to the solenoid field **H** times the coil cross-sectional area, **A**.

Into one of the coils is placed a tightly fitting long sample of the magnetic material. Typically, the sample is a glass tube filled with iron oxide. The sample is made long for two reasons. First, it simplifies matters if the demagnetizing fields of the long (cylindrical) sample are negligibly small and, second, the sample must extend considerably beyond the ends of the pick-up coil. The second coil, termed the "bucking" or cancellation coil, remains empty. The voltage induced in the filled coil is

$$E = -10^{-8} NA \frac{d}{dt}(H + 4\pi M) \tag{1.16}$$

Usually, the coil outputs are integrated electronically and displayed on the vertical axis of an oscilloscope. A signal proportional to the solenoid current and field, derived from a measuring resistor, appears on the horizontal axis. The resulting display is a $4\pi M$–H loop; if the bucking coil is not used, the display is of course, a B–H loop.

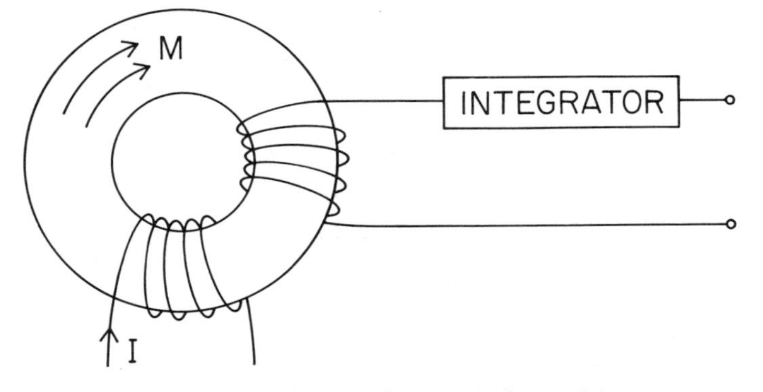

Fig. 1.10. A B–H looper (schematic).

The advantage of 60-Hz loopers are that they are simple, cheap, and fast in operation. They are ideally suited to such a task as incoming quality control of oxides at a magnetic tape factory. The disadvantages of 60-Hz loopers include the fact that their absolute precision is generally low. This low precision is related to the repetitive 60-Hz cycling of the magnetic material, which causes appreciable heating, and surprisingly, to the electronic difficulties which attend the design of low-frequency integrators. Generally the solenoid fields are limited to below 5000 Oe.

The toroidal B–H measurement technique shown in Figure 1.10 is generally used for magnetic head or transformer materials. The toroidal sample geometry is used because it completely avoids the demagnetizing field problems. A uniformly magnetized toroid can be regarded as a long cylinder folded onto itself and has zero demagnetizing fields because $\nabla \cdot \mathbf{M}$ is everywhere zero. Generally two coils are wound on the toroid. The primary coil carries the drive current and provides the magnetic field. Except at extremely high frequencies ($<$100 MHz), it is immaterial whether the coil is lumped or wound uniformly around the toroid's circumference. The field is given by the solenoid formula of Equation (1.2), where l is the average circumference of the toroid. The other, secondary, coil acts as the pick-up coil and is connected to an electrical integrator and an oscilloscope. The instrument displays B–H loops. The advantages of toroidal B–H loopers include cheapness, accuracy, and speed. B–H loopers, however, are not suitable for tape or disc magnetic materials because too much power is required to switch them at high frequencies.

The third type of instrument in general use, particularly in magnetic recording research, is the VSM. In a VSM, the time rate of change of magnetic flux in the pick-up coils is produced by mechanically vibrating the magnetized sample. The general arrangement is shown in Figure 1.11.

The magnetizing field in a VSM is usually provided by an electromagnet; typical maximum fields are 10,000 Oe. The magnetic field can only be changed slowly over periods of tens of seconds. The sample is suspended on a nonmagnetic rod, which vibrates vertically over an amplitude of perhaps one sixteenth of an inch at a frequency chosen to be incommensurate with 60 Hz. The pick-up coils are connected in push pull; that is, as the sample ascends, both the upper coils

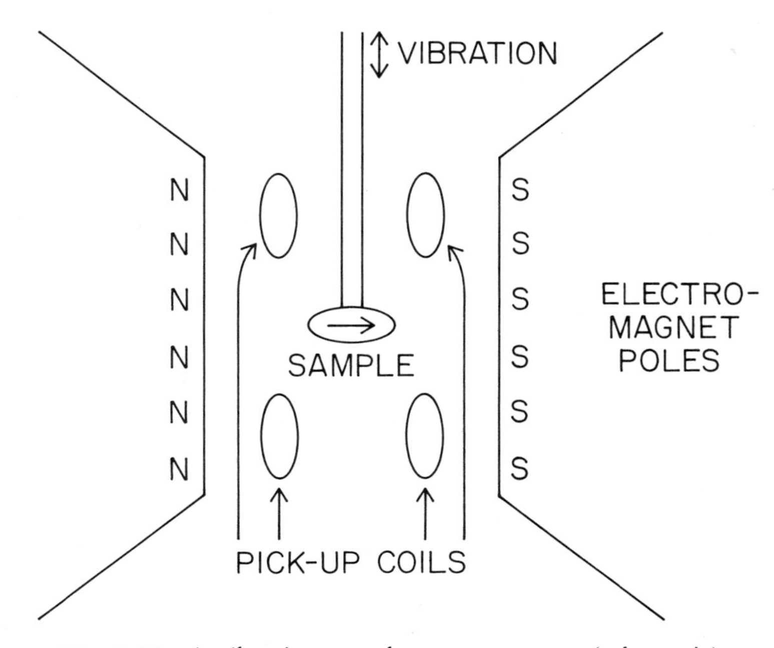

Fig. 1.11. A vibrating sample magnetometer (schematic).

and the lower coils produce the same electrical polarity. The ac voltage, which is proportional to the sample's magnetic moment, produced by these coils is taken to an extremely high gain lock-in amplifier. This lock-in amplifier is effectively an extremely narrow bandwidth amplifier. Typically, gain factors of 10^{+5} are achieved in bandwidths of 10^{-1} Hz centered on the vibrational frequency. The rectified output of the lock-in amplifier is applied to the vertical axis of an x–y chart recorder. With noise levels equivalent to 10^{-5} to 10^{-6} emu, satisfactory loops are achieved for samples with 10^{-3} to 10^{-4} emu of magnetic moment. A disc of recording tape about one quarter of an inch in diameter suffices.

Vibrating sample magnetometers have several disadvantages. They are, of course, expensive and slow. Since the coupling factor between the sample's magnetic moment and the pick-up coils cannot be calculated very accurately, calibration samples, usually made of nickel powder, must be used. The sample demagnetization factor cannot be always ignored. Very often spheres, which can be ground accurately, are used. Otherwise flat, discs of tape are used which have a negligible demagnetization factor within the plane of the disc. The greatest advantage of the VSM is that, by controlling the electromagnet field appropriately, any sequence of fields can be applied to the sample. One is able to navigate freely within the M–H plane, an ability which is denied in, for example, 60 Hz loopers, which are locked to eternal sinusoids.

Exercises

1. A dc current of 1 ampere flows into a 1000 turn solenoid that is 1 meter long. What is the internal field?
2. Give an expression (using e, h, and m) for, and the magnitude of, the Bohr magneton.
3. How many electrons does a neutral iron atom have, and what is its uncompensated electron spin moment?

4. What is the demagnetizing factor of a sphere?
5. What is the flux density, **B**, inside a uniformly magnetized sphere when it is placed in a region that is originally field free?
6. Give the definition of the magnetization, **M**.
7. What is the flux density, **B**, in an infinite flat plate, which is in a region that is originally field free, when it is magnetized uniformly at an angle, θ, to the plane of the plate?
8. What is the magnitude of the external field, **H**, of the plate in exercise 7?
9. What is the difference between the field, **H**, from a long filamentary current and a long filamentary line of poles?
10. What is the difference between an Amperian and a real current?
11. What is the magnetic field, **H**, midway between two long straight conductors carrying equal currents of the same polarity (the wires are 10 cm apart and carry 10 A)?
12. What are the only two sources of magnetic field, **H**?

Further Reading

Bozorth, Richard M. (1951). *Ferromagnetism.* Van Nostrand-Reinhold, Princeton, New Jersey.
Mee, C. D. (1964). *The Physics of Magnetic Recording.* North-Holland Publ., Amsterdam.
Mee, C. D., and Daniel, E., eds. (1986). *Magnetic Recording,* Vol. 1: *Technology.* McGraw-Hill, New York.
Smit, J., and Wijn, H. P. J. (1959). *Ferrites.* Wiley, New York.
Watson, J. K. (1980). *Applications of Magnetism.* Wiley, New York.
White, Robert M., ed. (1985). *Introduction to Magnetic Recording.* IEEE Press, New York.

Chapter 2

Hysteresis Loops and Multidomain, Single-domain, and Superparamagnetic Behavior

2.1 Introduction

Hysteresis results from irreversible changes that cause the dissipation, that is, the conversion to heat, of energy. Thus in magnetism, the area enclosed by a hysteresis loop is directly proportional to the energy taken from the driving magnetic field and converted into heat for each cycle around the loop. The specific irreversible phenomena involved are discussed in this chapter.

Many differing types of hysteresis loops are important in magnetic recording. Nevertheless, certain features of the loops are common and must be understood.

2.2 Major and Minor Loops

The typical appearance of a hysteresis loop is shown in Figure 2.1. The outer perimeter is the major loop and is the only unique feature. Within the loop exists an infinitude of differing minor loops. Only by applying a sufficiently high magnetic field can the memory of the previous fields be

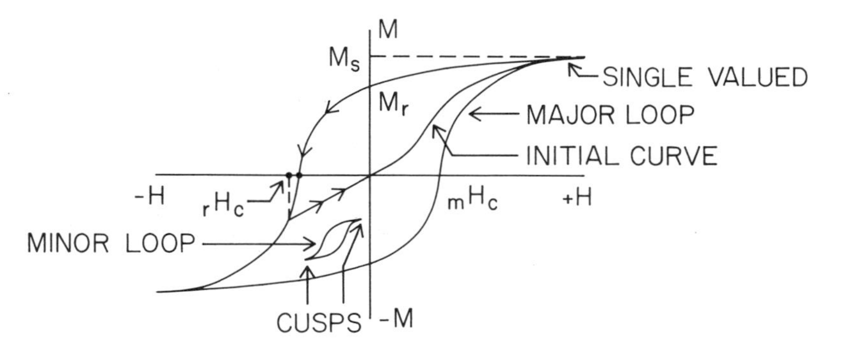

Fig. 2.1. *M–H* loop characteristics.

erased and the unique major loop be achieved. The specific test for attainment of the major loop is that the loop must close, that is, become single valued, over a range of applied fields. It is not enough that the loop tips form cusps; closure is needed.

Given that the major loop has been achieved, several unique, repeatable characteristics can be determined. The maximum magnetization is called the saturation magnetization, M_s. The magnetization remaining when the field H is reduced to zero is called the remanent magnetization, M_r. The ratio $M_r : M_s$ is called the squareness ratio. The field required to reduce the magnetization to zero is called the intrinsic coercive force ${}_mH_c$. If the negative field is increased further, the remanent coercive force ${}_rH_c$ is reached where, upon reducing the field to zero, zero remanent magnetization remains. Obviously the magnitude of ${}_rH_c$ is greater than that of ${}_mH_c$.

Starting from zero remanent magnetization, an increasing positive field causes the so-called initial, or virgin, curve to be traced out. The slope of this curve is called the susceptibility, χ. For small fields, the initial curve is linear, so that $M = \chi H$.

Similar characteristics are shown in the B–H loops. Thus there is the saturation flux density, B_s, the remanent flux density, B_r, the squareness ratio $B_r : B_s$, the technical coercive force ${}_bH_c$, and the initial slope, which is called the permeability μ. Since $B = H + 4\pi M$, clearly $\mu = 1 + 4\pi\chi$. Note also that B–H loops may be transformed into M–H loops, and vice-versa, by graphical or numerical manipulations because they contain precisely the same information about the magnetic sample. While the technical ($B = 0$) coercive force magnitude is less than that of the intrinsic ($M = 0$) coercive force, the areas within both loops are identical.

2.3 Shearing and Unshearing Operations

Whenever a loop, say the M–H loop, has been measured on a sample for which the demagnetization factor is not negligibly small, the loop needs correcting for this fact. This is because the total magnetic field experienced by the sample is not only the field provided by the electromagnet or solenoid but also includes the demagnetizing field.

$$H_t = H_s - NM \tag{2.1}$$

For example, if the VSM sample shape is spherical, $N = 4\pi/3$. As is shown in Figure 2.2, the effect of the demagnetization field is to shear, or distort, the loop. The appearance

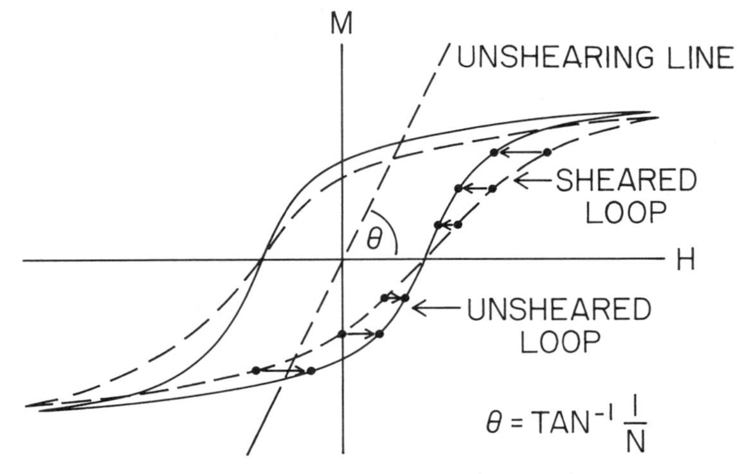

Fig. 2.2. Sheared and unsheared M–H loops.

is as though the loop tips have been pulled in opposite directions along the H axis. The required correction operation, termed unshearing, can be conducted graphically or numerically.

Graphically, an unshearing line of slope arctangent $1/N$ is constructed, as shown in the figure. Then all points on the sheared loop are moved horizontally as indicated. In this manner, the horizontal axis is corrected to become the total field. Note the unshearing does not change the coercive force because, when $M = 0$, the demagnetizing field is zero also. Moreover, the loop area is conserved during these operations.

2.4 Remanent Magnetization Loops

The third kind of loop is obtained by plotting the remanent magnetization versus field. In this case the magnetic field is applied and then turned off repeatedly in order to measure the magnetization remaining in zero field. Obviously this can not be achieved with sine wave drive instruments. Remanent loops are generally measured using VSM and are somewhat time consuming. Nevertheless, for tape and disc magnetic materials, it is an appropriate thing to measure because, after being written by the magnetic field, tapes and discs are reproduced in an almost zero field condition. Note that M_r–H loops contain different information than do M–H and B–H loops and cannot, therefore, be derived from them.

2.5 Anhysteresis

If a magnetic sample is subjected to two coaxial magnetic fields, where one is of larger but decreasing amplitude and alternating polarity and the other is of smaller amplitude and fixed polarity, the phenomenon of anhysteresis results. In this case, all the energy required to switch the magnetization back and forth is provided by the ac field. The final direction and magnitude of the magnetization is determined by the dc field. A plot of the remanent magnetization obtained when first the ac field and then the dc field are reduced to zero is called the anhysteretic curve and it is well approximated by the error function (i.e., the integral of a Gaussian curve) for nearly all magnetic materials. Interest centers on the anhysteretic curve because the center portion of the curve is linear and makes possible distortionless recording in ac-biased magnetic recorders. Usually, Audio and Instrumentation recorders are ac biased; they are discussed in detail in Chapter 8.

2.6 Multidomain Behavior

In the rest of this chapter, the M–H loop behavior of samples of decreasing size is followed. For samples of millimeter, micron, and submicron size, the magnetic behavior is described as multidomain, single domain, and superparamagnetic, respectively.

These behaviors have their origin in the minimization of the total energy of the system. Accordingly, it is necessary to begin by defining magnetic energies.

The magnetostatic, or Zeeman, energy of a sample in an external magnetic field is given by

$$E_{\mathrm{m}} = -\int \mathbf{M} \cdot \mathbf{H}\, dv \tag{2.2}$$

and, accordingly, like a compass needle, the magnetization tends to align parallel to magnetic fields. The self-, or demagnetization, energy is

$$E_{\mathrm{d}} = -1/2 \int \mathbf{M} \cdot \mathbf{H}_{\mathrm{d}}\, dv \tag{2.3}$$

where the factor one-half arises for just the same reason it appears in the familiar expressions for the energy of inductors and capacitors, $\frac{1}{2}LI^2$ and $\frac{1}{2}CV^2$, respectively. For ellipsoids of revolution, uniformly magnetized along a principal axis, $E_{\mathrm{d}} = \frac{1}{2}NM^2$ because $H_{\mathrm{d}} = -NM$, and the only way to reduce E_{d} is for the sample to reduce M or demagnetize. In Equations 2.2 and 2.3, M and H are vectors multiplied by the scalar, or inner product.

Consider now an ellipsoidal sample of the polycrystalline iron of length, say, one centimeter. In order to reduce the demagnetizing energy, the magnetization breaks up into millions of domains. Within each domain the magnetization is uniform, nearly parallel to an easy axis, and has magnitude M_s. Between the domains, the magnetization changes direction by means of a domain wall. Domain walls are typically 0.1 μm (micron) thick and have usually, but not always, a divergence-free rotation of the magnetization, which, of course, produces no magnetic poles. Thus the component of magnetization normal to the wall in the adjacent domains is continuous; walls bisect the angle between magnetizations. The energy per unit area of an unperturbed domain wall is

$$E_w = \sqrt{AK} \tag{2.4}$$

where E_w is in ergs per square centimeter, A is the exchange constant (erg/cm^{-1}), and K is the magnetocrystalline energy constant (erg/cm^3). The domain wall energy is a trade-off between exchange, which tries to force adjacent spins to be parallel, and magnetocrystalline anisotropy, which favors magnetization along the easy axes. Domain wall energy is analogous to surface tension.

Figure 2.3 shows the total energy of the sample in zero external field as a function of its magnetization. Note the basic parabolic shape due to the self-, or demagnetizing, energy. In addition, the total energy varies from point to point due to the changes in domain wall area and energy. As the magnetization increases, the domain walls move through

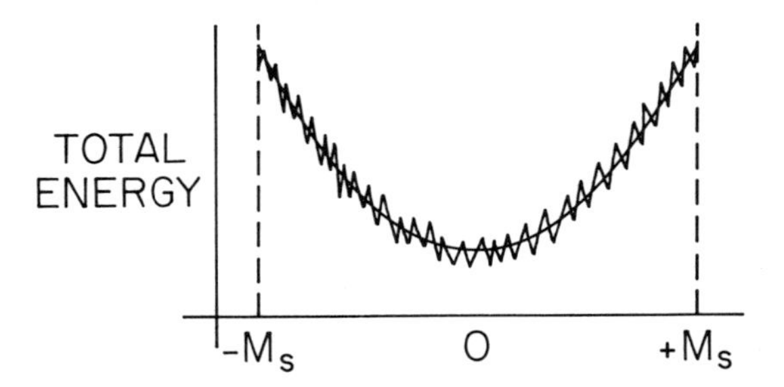

Fig. 2.3. Magnetic energy versus magnetization.

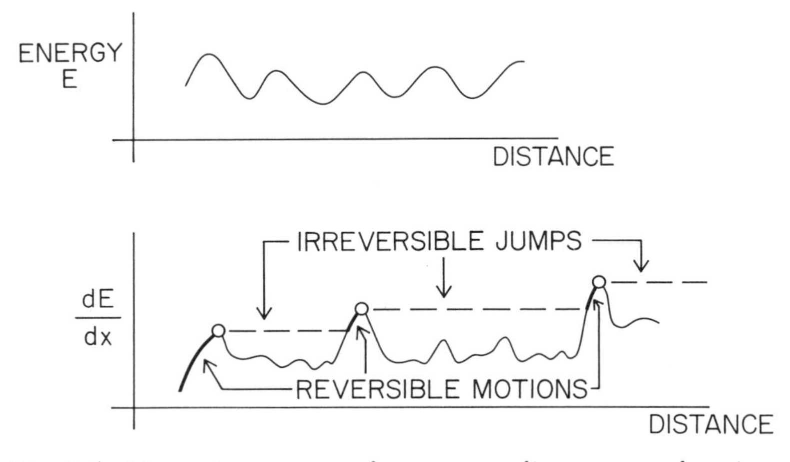

Fig. 2.4. Magnetic energy and energy gradient versus domain-wall position showing irreversible and reversible motions.

the sample in a way that increases the volume of favorably oriented domains and decreases the volume of other domains. The free passage of the domain walls is impeded by a variety of obstacles such as foreign inclusions, grain boundaries, and crystal dislocations. Accordingly, the domain wall area and energy change from point to point. A good analogue is the motion of soap film bubbles through the meshes of a nylon or wire pan scrubber. In Figure 2.4 is shown the energy and rate of change of the wall energy versus magnetization, or distance, when the magnetization is close to zero.

Suppose that, in response to an external field H_c, the wall moves a distance dx and the magnetization increases $2M_s\,dx$. The change in energy per unit width is by Equation 2.2,

$$dE = (2M_s dx)\,H_c \tag{2.5}$$

or

$$H_c = \frac{1}{2M_s}\frac{dE}{dx} \tag{2.6}$$

showing that the coercive force is proportional to the spatial derivative of the wall energy.

The motion of the domain wall in response to a slowly increasing field around H_c also is shown in Figure 2.4. First the wall moves reversibly, in equilibrium, until it reaches a local extremum, or peak, in the derivative plot. During this motion, the magnetic system is simply storing energy. Thereafter, upon a further slight increase in field, the domain wall breaks free and moves rapidly and irreversibly to the side of the next peak. During the irreversible jump, magnetic energy is dissipated or converted into heat. The entire hysteresis loop is made up of such elemental jumps as is indicated in Figure 2.5. These domain-wall jumps were first detected by Barkhausen; it is these jumps that are responsible for Barkhausen noise.

Note that since the self-, or demagnetizing, energy is zero when $M = 0$, the intrinsic coercive force, ${}_mH_c$, is dependent only on the domain wall energy variations. Typically, in materials used for magnetic heads and transformers, $0.1 \text{ Oe} < {}_mH_c < 10 \text{ Oe}$. In order to reduce the coercive force, increase the susceptibility, χ, and decrease the hysteresis loop area, several steps are taken in the preparation of such so-called soft magnetic materials. Examples include annealing to remove mechanical strains and to eliminate voids, the use of pure starting materials, and the avoidance of preparation conditions which produce second phases and other inclusions.

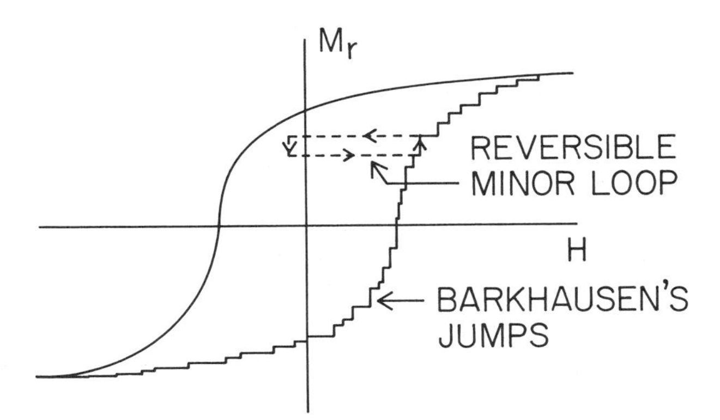

Fig. 2.5. Remanent magnetization loop showing Barkhausen's jumps.

2.7 Single-domain Behavior

Now suppose that the sample is reduced in size so that the length becomes somewhat less than 1 μm. Figure 2.6 shows, schematically, the variations with size of the domain wall and the demagnetizing energies. Note that whereas the wall energy varies with area, the self-energy depends on volume. At the so-called critical size, the demagnetizing energy falls below the wall energy, and below that sample size it is no longer energetically favorable for the magnetization to subdivide into the multidomain state. Below a length of about 1 μm, the magnetic sample, or particle, is single-domain in zero external field.

Many calculations have been made to elucidate the hysteretic properties of such single-domain particles. The calculations separate into those in which the particle reverses its magnetization in the single-domain state, termed coherent rotation, and those in which the magnetization does not remain uniformly parallel during the switch, called incoherent rotation. It is important to realize that the sample, or particle, remains physically fixed in these rotations; only the magnetization changes direction. The problem of calculating the reversal behavior of a single-domain particle is in many ways analogous to the problem of computing the load-bearing strength of a column. In each case, the essence of the prob-

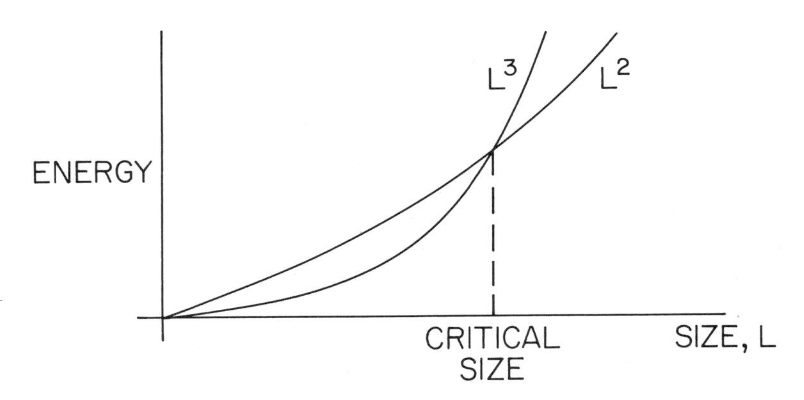

Fig. 2.6. Magnetostatic and domain-wall energies versus particle size.

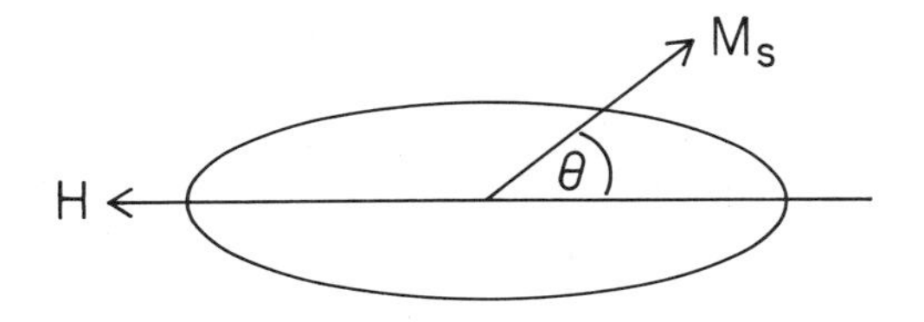

Fig. 2.7. Coherent rotation in a particle.

lem is that of determining the manner, called the mode, by which the system responds. Once given that a column fails by certain bending mode, it is easy to compute the strength. Determining the mode is the difficult part. In single-domain particles, one seeks the lowest energy eigenmodes of reversal. For ellipsoids of revolution, these have been found to be the incoherent modes, curling and buckling and coherent rotation. In this chapter, for the sake of brevity, only coherent rotation and an approximate incoherent mode called fanning are described.

In the Stoner–Wohlfarth model, shown in Figure 2.7, the energy density,

$$E = \frac{M_s^2}{2}(N_\perp \sin^2 \theta + N_\parallel \cos^2 \theta) + MH \cos \theta \quad (2.7)$$

where E is the energy density (erg/cm^3), $N_\perp$ and $N_\parallel$ are the demagnetization factors perpendicular to and parallel to the particle axis, θ is the magnetization angle, and H is the (negative) magnetic field.

Next the equilibrium magnetization angle θ is found by the usual procedure: differentiate E with respect to θ and set the differential equal to zero. From this, one concludes that only two stable magnetization angles are possible, $\theta = 0°$ and 180°. The critical field, sometimes called the nucleation field, at which the magnetization reverses is

$$H_n = (N_\perp - N_\parallel)M_s \quad (2.8)$$

For long ellipsoids, say greater than 5 : 1 axial ratio, $N_\perp - N_\parallel = 2\pi$ to a good approximation. Accordingly, the Stoner–Wohlfarth nucleation field, which is here also the coercive force because of the alignment of the field and particle's long

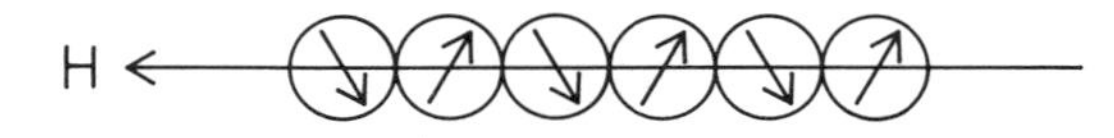

Fig. 2.8. The chain-of-spheres fanning model.

axis, is $2\pi M_s$. For gamma-ferric oxide and iron, the expected coercive forces are 2350 Oe and 10,500 Oe, respectively. That these values are about a factor of eight higher than is experimentally observed may be taken to indicate that the coherent mode does not actually occur in physical reality.

The incoherent model considered here is shown in Figure 2.8. In the fanning model, the particle is imagined to be subdivided into a number of equal volume spheres; for this reason the model is often called the chain-of-spheres model. In the analysis of this model, the magnetization within each sphere rotates in alternate directions, hence the name fanning. The nucleation, or coercive, field is

$$H_c \approx M_s \tag{2.9}$$

Thus for gamma-ferric oxide and iron, the expected values are 370 Oe and 1700 Oe, respectively, which is in fairly good accord with measurements.

The good agreement is indicative that a reversal mode similar to fanning actually takes place. It may be speculated that the incoherent reversal is promoted by such nonuniformities of the particles as voids, grain boundaries, dislocations, mechanical strains, and their irregular shape.

2.8 Superparamagnetism

Now consider an even smaller particle of length under 0.1 μm. At this small size, even though the particle is uniformly magnetized, thermal energy is sufficient to switch the particle's magnetization. Suppose an assembly of such particles is magnetized in the same direction by the application of a suitably large magnetic field, which is then turned off. The magnetization spontaneously decays as shown in Figure 2.9,

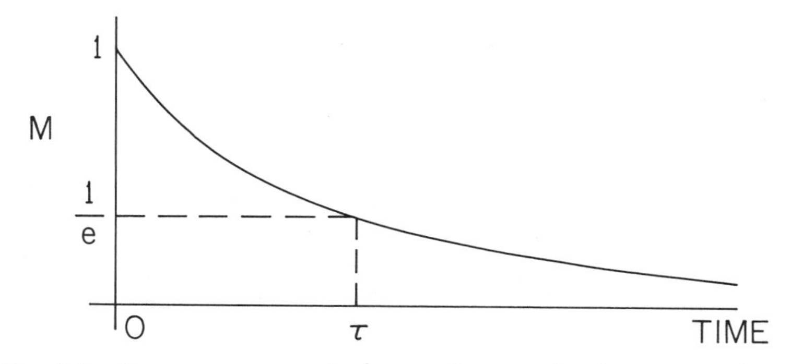

Fig. 2.9. Superparamagnetic decay of magnetization versus time.

according to

$$M(t) = M(0) \exp\left(-\frac{t}{\tau}\right) \tag{2.10}$$

where $M(t)$ is magnetization at time t, $M(0)$ is the initial magnetization, t is the time in seconds, and τ is the superparamagnetic relaxation time.

The relaxation time is given by

$$\tau = 10^{-9} \exp\left(\frac{0.5\, M_s H_c v}{kT}\right) \tag{2.11}$$

where k is Boltzmann's constant, 1.38×10^{-16} and T is the absolute temperature. The energy required to switch a particle is $0.5\, M_s H_c v$. Every degree of freedom of a system in thermal equilibrium at an absolute temperature T has thermal energy equal to $0.5kT$. When the switching energy is 50 times thermal energy, that is,

$$0.5\, M_s H_c v = 25kT \tag{2.12}$$

then the relaxation time is 100 seconds. Because of the exponential form of equation 2.11, the relaxation time changes very rapidly with changing particle size. For example, if the dimensions of a 100 second particle are increased by only 20%, the volume increases by 1.2 cubed and the relaxation time becomes over 100 years. Practically, this means that, when a group of particles has a wide distribution of sizes, there exists, at any temperature, a clear-cut boundary between those which are stable single-domain particles and those which are thermal idiots. Particles rather larger than that boundary give rise to the phenomenon of print-through in reels of tape.

When the M–H behavior of an assembly of very small superparamagnetic particles is measured, a single-valued characteristic is found. For every value of the applied field, there is but one thermal equilibrium value of the magnetization,

$$M(H) = M_s \tanh\left(\frac{M_s H v}{kT}\right) \tag{2.13}$$

The situation is related to that of anhysteresis, where the ac field provides the necessary switching energy. In superparamagnetism, thermal energy does the work. Note that unlike anhysteresis, however, a superparamagnetic sample has zero long-term remanence.

Exercises

1. What value of the coercive force is predicted by the chain-of-spheres fanning mode of reversal of a single-domain particle: $2\pi M_s$, M_s, or does it depend on size?
2. Give the criterion for 100 second relaxation time superparamagnetism.
3. What is the volume of 100 second superparamagnetic γ-Fe_2O_3 ($M_s = 350$ emu/cm^3, $H_c = 300$ Oe) particles at room temperature (300 K)?
4. What is the flux density, **B**, in an infinite flat plate, uniformly magnetized out of the plane, made of magnetic material with $H_c = 500$ Oe and $4\pi M_s = 10{,}000$ G when it is immersed in a uniform field of 1000 Oe normal to the plate?
5. Put the $B = 0$, $M = 0$, $M_r = 0$ coercive forces in numerical order.

6. When cycling a magnetic material around its hysteresis loop, what phenomena cause it to convert magnetic energy into heat?
7. What is the test for a major loop?
8. What is the relationship between the permeability and the susceptibility of a magnetic material?
9. Why is the coercive force unchanged when unshearing an M–H loop?
10. What is the origin of magnetic poles?
11. Why do domains form in large magnets?
12. What coercive force would you expect a long single-domain metallic iron particle to have?

Further Reading

Bertram, H. N. (1968). Monte Carlo calculation of magnetic anhysteresis. *J. Phys.* (Paris) **32**, 684–685.

Kittel, Charles (1949). Physical theory of ferromagnetic domains. *Rev. Mod. Phys.* **21**, 541–583.

Smit, Jan (1971). *Magnetic Properties of Materials.* McGraw-Hill, New York.

Chapter 3

Magnetic Recording Media

3.1 Introduction

There exists a wide variety of magnetic recording media and heads. Particulate media include gamma-ferric oxide, cobalt-modified gamma-ferric oxide, chromium dioxide, metallic iron, and barium ferrite. Continuous media include cobalt–phosporus, nickel–cobalt–phosporous, and cobalt–chromium thin metallic films.

3.2 Gamma-Ferric Oxide

Gamma-ferric oxide, γ-Fe_2O_3, has been used in commercial tape manufacturing since 1937 and remains one of the principal recording materials used today.

Figure 3.1 shows a typical γ-Fe_2O_3 particle as is used in both tape and discs. It has the following properties: length, 0.25–0.75 μm; width, 0.05–0.15 μm; length-to-width ratio, 5 to 10 : 1; void volume, 5–10%; saturation magnetization, M_s = 370 emu/cm^3; specific saturation magnetization, σ_s = 76 emu/gm; specific gravity, ρ = 4.9 gm/cm^3; coercive force, ${}_mH_c$ = 300 Oe; Curie temperature, T_c = 600° C (by extrapolation); with a transformation temperature of about 400° C.

Gamma-ferric oxide is manufactured as a high-price specialty item by companies which make coloring pigments for the paint industry. For example, hematite (α-Fe_2O_3) and barium ferrite ($BaFe_{12}O_{19}$) are red and green pigments, respectively. Gamma-ferric oxide itself is used also as a brown

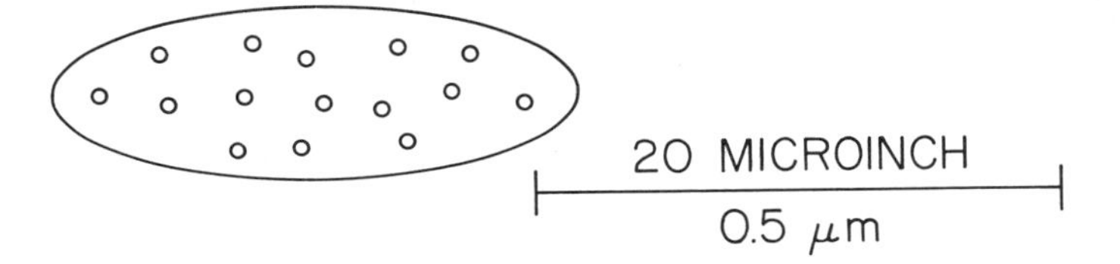

Fig. 3.1. A typical γ-Fe_2O_3 particle.

pigment. Even when made for the recording industry, γ-Fe_2O_3 is manufactured on a very large scale, with lots of 10,000 lb (five short tons) being the norm. The majority of γ-Fe_2O_3 is made following the chemical process outlined below.

Step one. A slurry of minute colloidal particles of $Fe(OH)_2$ is prepared, for example, by adding ammonia (NH_4OH) to a solution of iron sulfate ($FeSO_4$). Although these particles are only 50–100 Å (0.005–0.01 μm) in length, their length-to-width ratio is about the same as that of the final product. All subsequent steps in the process maintain the original shape of the slurry particles.

Step two. The slurry is introduced to a solution of iron sulphate ($FeSO_4$) through which air is bubbled. The container may well be tens of feet in dimension, and the solution is very often "pickling fluid" from a steel rolling mill. Pickling fluid is produced in the process of washing steel strip with sulphuric acid (H_2SO_4) in order to remove iron oxide scale. Over a period of days, elongated particles of goethite or alpha-ferric oxide monohydrate (α-$Fe_2O_3 \cdot H_2O$) grow. The growth is usually monitored by observing the exact color of the particles. The trained human eye can easily detect hue changes which correspond to a change in length of only 50 Å in a 5000 Å particle. The product of this step is called yellow oxide, and the final product's particle size is now determined. The specific gravity of the yellow is only 4.0 gm/cm^3.

Step three. The yellow is now heated, in air, to about 200° C. The ensuing dehydration produces hematite (α-Fe_2O_3), commonly called jeweler's rouge. Because the specific gravity has now increased to 5.0 gm/cm^3, but the overall particle size has not changed, inevitably voids appear. These voids, which may promote the fanning mode of switching, remain in the particle in subsequent processing. Red hematite is antiferromagnetic and has an rhombohedral crystal habit.

Step four. The hematite is now chemically reduced in hydrogen (H_2), carbon monoxide (CO), or mixtures thereof, at temperatures of 250–350° C to produce magnetite (Fe_3O_4). Magnetite is ferrimagnetic with a specific saturation magnetization of 90 emu/gm and has a cubic crystal habit. Unfortunately, magnetite cannot be used in recording media because it slowly oxidizes in air. A continuous range of chemical compounds between iron and gamma-ferric oxide exists and tapes made with magnetite do not have stable magnetic properties.

Step five. The magnetite is next oxidized in moist air at a temperature of 200–250° C. The oxidation reaction is exothermic and consequently great care must be taken to maintain and control the precise oxidation temperature. The presence of water vapor promotes a lower oxidation temperature. The result is gamma-ferric oxide, which has several other names: maghemite, ferroso-feric, and brown oxide. It is the permanent magnet material of the lodestone, and as the principal iron ore, is probably the most abundant magnetic material extant. Gamma-ferric oxide particles produced in this manner are chemically stable at all normal environmental conditions. When heated to above 400° C, however, the cubic crystal transforms to hematite. Alpha-ferric oxide and γ-Fe_2O_3 are allotropes; that is, they are the same chemical compound with different crystal structures. The heating of γ-Fe_2O_3 facilitates the very small change in atomic positions that causes the transformation from cubic to rhombohedral. Hematite is the lower energy, more stable phase, and the process can only be reversed by cycling through steps four and five again.

The gamma-ferric oxide particles are packed in tape at about 40–50% by volume by coating techniques that have much in common with painting. A dispersion is made which contains oxide, binder, and solvent. The binder system is, almost universally, a polyester-urethane polymer. The dispersion is coated on a web of base film by a number of techniques: knife, roll, and gravure coating. The base film is, again almost universally, polyethylene-ptherthalate (Mylar, Celanar, Estar) and is about 25 μm (1 mil) thick and several feet wide. Coating speeds of 300–400 feet per minute are common. Over 600 square miles of recording media was coated in 1984.

The coated tape is next subjected to a magnetic orienting field, typically of magnitude 1000 Oe, in order to align physically the particles before the solvent evaporates and the "paint dries." Squareness ratios of 0.75 to 0.85 are achieved, with the particles more or less evenly distributed in a ± 30 degree solid angle cone around the orientation direction. Since the coercive force of a particle depends on the angle between the applied field and the particle's long axis, the process of orienting tapes greatly reduces the distribution of particle coercivities. This makes possible better recording at high density, as will be discussed in a later chapter.

After orientation, the coating is dried by passing it through a heated oven. During this phase the coating decreases in thickness by about a factor of five. Thereafter, the tape is usually calendered, that is squeezed between rubber, plastic, or steel rollers in order to "densify" the coating (that is, to remove air bubbles) and to impart as smooth a surface as is possible. Smooth surfaces are critical to high density recording. The finished tape is then slit to the proper width and reeled up.

In the manufacture of flexible, or floppy, discs, most of the same processes are followed, with the exception that orientation is not required or desired. In order to remove the slight orientation caused by hyrodynamic forces in the coating process, normally a deorienting field is applied transverse to the web motion.

For computer rigid, or hard, discs, the dispersion yields only a 20–25% pigment volume concentration (pvc), and the binder system is usually one of the epoxy resins chosen for its mechanical hardness. The dispersion is sprayed onto the spinning aluminum alloy disc and is followed by orientation, drying, and polishing.

A γ-Fe_2O_3 tape with a pvc of 45% and a squareness ratio of 0.8 has maximum remanent magnetization given by $4\pi M_r = (4\pi)(370)(0.45)(0.8) = 1675$ G. Rigid discs, on the other hand, typically have only 800–900 G due to their lower pvc's and poorer particle orientations.

3.3 Cobalt Modified γ-Fe_2O_3

In order to increase the coercive force of γ-Fe_2O_3, the particles produced in step five above may be subjected to further processing. These additional steps add cobalt atoms to the surface of the gamma-ferric oxide particle, forming a layer, perhaps only 30 Å thick, of a compound close to cobalt ferrite ($CoFe_2O_4$) in composition.

Step six. The gamma-ferric oxide particles are placed in an aqueous solution of, say, cobalt chloride ($CoCl_2$) and sodium hydroxide (NaOH) and held at a temperature of about 80° C for several hours.

Step seven. The particles are dried at a temperature of 100–150° C, which is too low to permit significant diffusion of cobalt atoms to the interior of the particle.

By adding surface material of high magnetocrystalline anisotropy, the lower energy fanning modes are suppressed and coercive forces as high as 800–1000 Oe are achieved. Since 1976, most professional and consumer video tape has contained such particles, with ${}_{m}H_c = 600$ Oe. The computer disc world has been slow to adopt the technique; only in 1984 did the so-called "chocolate" (colored) discs appear commercially.

A notable feature of these surface-modified particles is that the increase in ${}_{m}H_c$ is relatively independent of temperature (about 2 Oe per degree centigrade). If they are heated to about 250–300° C, the cobalt atoms diffuse deep into the particle, and though the increased coercivity is retained, the increase becomes highly temperature dependent (about 10

Oe per degree centigrade). Such bulk-diffused cobalt gamma-ferric oxide particles are not used in magnetic recorders for this reason.

3.4 Chromium Dioxide

Before the surface-modification process was discovered in Japan in the early 1970s, the need for particles with higher coercivity than that of pure γ-Fe_2O_3 led to the introduction of chromium dioxide. It is still used, to a limited extent, in consumer video and half-inch computer tapes. Chromium dioxide (CrO_2) is grown by hydrothermal processes at high temperatures (500° C) and high pressures (500 atm). Consequently, it is relatively expensive compared with iron oxides. It is a ferromagnet with the following properties: M_s = 490 emu/cm^3; $_mH_c$ = 450 Oe, which can be increased by the addition of other elements; and a Curie temperature of only 128° C. Chromium dioxide tapes are made in the usual way and have somewhat higher remanent magnetization, not only because the M_s is higher, but also because the particles can be oriented rather more perfectly. Squareness ratios of 0.9 have been achieved, which is attributed to the "clean" particle morphology; that is, the particles are free of the dendritic growths that were found on the older iron oxides.

The lower Curie temperature has encouraged work in thermal contact duplication of video tapes. Here, a mirror image "master" recording made on iron oxide tape is reeled, or otherwise placed in close contact, coating side to coating side with chromium dioxide "slave" tape. The pair is heated to above the CrO_2's Curie temperature and allowed to cool. A process, called thermomagnetic magnetization, which is very similar to that of anhysteretic magnetization, occurs as the CrO_2 particles go through their superparamagnetic temperature range just below the Curie temperature. When the CrO_2 particles reach their blocking temperature and become again thermally stable single-domain particles, the magnitude and direction of their remanent magnetization has been set by the fringing fields from the master tape recording. As the CrO_2 tape cools to room temperature, its magnetization increases further, yielding a very high value of thermoremanent susceptibility,

$$\chi_T = \chi_A \left(\frac{M_s(\mathrm{BT})}{M_s(\mathrm{RT})} \right) \tag{3.1}$$

where $M_s(\mathrm{BT})$ and $M_s(\mathrm{RT})$ are the saturation magnetizations at the blocking and room temperatures, respectively, and χ_A is the anhysteretic susceptibility.

Despite the intuitive appeal of the process as a means of inexpensive mass duplication of video tapes, little commercial application has occurred. This is mainly due to the fact that the process does not copy high density information well.

3.5 Metallic Iron Particles

In order to obtain even higher coercive forces and simultaneously increase the magnetization substantially, iron particles are being used more often. Iron particles may be manufactured by further processing of conventional iron oxide particles.

Step six. Gamma-ferric oxide particles are soaked in an aqueous solution of, for example, tin chloride ($SnCl_2$) and are then dried at low temperatures.

Step seven. The coated particles are chemically reduced to metallic iron with hydrogen at temperatures close to 300° C. Without a surface coating of tin, or one of about a dozen other elements, the reduction is accompanied by pronounced sintering, that is, the bonding together by diffusion of the particles, rendering them useless. With the proper coating, however, the reduction is pseudomorphic and the particles retain their individual shapes.

Step eight. The reduced iron particles are exposed to a carefully controlled atmosphere of oxygen, nitrogen, or air. At low temperatures, this produces a surface layer of iron oxide or nitride and permits the now-protected, or inhibited, particles to be exposed to air. Without an inhibiting process, metallic iron particles are intensely pypophoric, igniting spontaneously in air. Further protection from oxidation is provided by the polymeric binder system used in the tape coating.

Typical properties of the iron particles used today in metal-particle (MP) consumer video tapes are M_s = 160–175 emu/cm^3 and $_mH_c$ = 1450 Oe. The tape remanence is approximately 3000 to 3500 G and the hysteresis loop area is, therefore, one order of magnitude greater than that of the original γ-Fe_2O_3 tapes.

3.6 Metallic Films

Many attempts are being made to use thin metallic film as recording media. Because metallic films have such high remanent magnetizations, coatings as thin as a few hundred angstroms yield outstanding recording performances. The principal disadvantages of thin metallic films concern wear, friction, and corrosion problems. Whereas in particular media, lubricants can be incorporated in the plastic binder system, in thin films the lubricant can only be applied to the surface. Moreover, since most magnetic metallic films are more chemically reactive than the particulate oxides, overcoat films designed to inhibit corrosion are needed. Over the last 25 years the following overcoat films have been tried without achieving complete success: rhodium, silicon dioxide, graphite, and fluorinated hydrocarbon oils. In metal-evaporated (ME) consumer video tapes, cobalt–nickel alloy is electron-beam evaporated at an oblique angle onto a moving web of base film. The alloy forms elongated columns which are inclined out of the plane of the tape. A consequence of this is that the recording performance differs slightly for the two directions of tape motion. Only wear and corrosion problems have prevented this tape from being adopted as the standard for the future helical scan, consumer video and audio recorders.

There are several types of computer, rigid disc, thin metallic films being investigated. Most are deposited on a relatively thick layer of superparamagnetic nickel–phosphorus (Ni–P), which is chemically plated on an aluminum disc. Nickel–phosphorus is used as a so-called subbing layer because it covers defects in the aluminum and because it is easy to polish properly. An electroplated cobalt–nickel–phosphorous, chemically plated cobalt–phosphorous, or vacuum sputtered cobalt–nickel–phosphorous alloy magnetic recording layer is then deposited, followed by some kind of protective overcoat. Typical magnetic properties are M_s = 1000 emu/cm^3 and $_mH_c$ = 750 Oe. In all these films, the easy axes of magnetization are distributed isotropically in the plane of the film, and consequently, they are called longitudinal or horizontal media.

Over a decade of development in Japan has been devoted to perpendicular or vertical thin media. In these media, the magnetization and easy axes are normal to the plane of the film. It has been widely believed that this perpendicular orientation should yield better high density recording than that attained with longitudinal orientation. Unfortunately, however, significant differences have not been found experimentally.

The usual material is a sputtered film of cobalt–chromium, which forms naturally into columns of about 1000 Å diameter oriented normal to the plane of the disc. Provided the out-of-plane demagnetizing field ($4\pi M_s$) is less than the magnetocrystalline anisotropy field ($2K/M_s$), the remanent magnetization can be perpendicular; this condition is achieved by making the chromium content greater than about 14 atomic percent. Below this percentage, longitudinal Co–Cr films are found.

In order to obtain the potential advantages of perpendicular recording without the disadvantages of metallic films, a recent trend is to attempt to orient barium ferrite ($BaFe_{12}O_{19}$) particles vertically in a conventional plastic binder system. Here again, however, not much difference in recording performance between the two orientations has yet been found.

Exercises

1. What is the $4\pi M_r$ of a γ-Fe_2O_3 tape, packed at 40%, having a 0.8 remanence ratio?
2. What superparamagnetic lifetime follows if the switching energy of a particle is $30kT$?
3. What is the external field, H, of a uniformly magnetized toroid?
4. What is the fundamental difference between the information contained in a B–H loop and an M–H loop?
5. What is the chemical difference between γ-Fe_2O_3 and Fe_3O_4?
6. What is the chemical difference between α-Fe_2O_3 and γ-Fe_2O_3?
7. If a uniformly magnetized body of volume V has 1000 emu of magnetic moment, what is the value of its magnetization, M?
8. What physical features of particles are thought to promote the lower energy fanning modes of reversal?
9. Under what conditions do ellipsoids of revolution have diagonal demagnetization tensors?
10. What is the sum of the elements of a diagonal demagnetization tensor in cgs–emu?
11. Which electron shell of iron contains the uncompensated spins?
12. Why does an antiferromagnet have almost zero moment?

Further Reading

Bate, G. (1981). Recent developments in magnetic recording materials. *J. Appl. Phys.* **52**, 2447–2452.

Iwasaki, Shun-Ichi (1984). Perpendicular magnetic recording—evolution and future. *IEEE Trans. Mag.* **20,** 657–668.

Mallinson, John C. (1976). Tutorial review of magnetic recording. *Proc. IEEE* **64,** 196–208.

Mallinson, John C., and Bertram, H. Neal (1984). A theoretical and experimental comparison of the longitudinal and vertical modes of magnetic recording. *IEEE Trans. Mag.* **20,** 461–467.

Mallinson, John C., *et al.* (1971). A theory of contact printing. *IEEE Trans. Mag.* **7,** 524–527.

Mee, C. D. (1964). *The Physics of Magnetic Recording.* North-Holland Publ., Amsterdam.

Chapter 4

Magnetic Recording Heads

4.1 Introduction

The structure of all magnetic heads used commercially is the same: they are all ring structures with a small gap. The ring parts are made of metallic alloys such as permalloy (Ni–Fe), alfesil (Al–Fe–Si), and cobalt–zirconum (Co–Zi), or from magnetic oxide compositions such as nickel–zinc and manganese–zinc ferrites. The gap length is generally determined by a sputtered layer of silicon dioxide, which is subsequently glass bonded by melting. Very often the same head is used for both writing or recording and reading or reproducing; this is the case in video recorders and computer disc recorders. The saturation flux density, B_s, and the gap length determine the maximum fringing field attainable above the gap and, therefore, the maximum coercivity of the recording medium. As a reproducing head, the overall size and the gap length determine the maximum and minimum wavelengths, respectively. In this chapter, the principal emphasis will be on the mathematical analysis of the inductance, efficiency, and fringing field shape of heads.

4.2 Magnetic Material Properties

Figure 4.1 shows two B–H loops with different minor loops. In the case of a writing head, the write current in the head coil drives the material to high flux densities (typically 1000–2000 G) in order to provide sufficiently high fringing

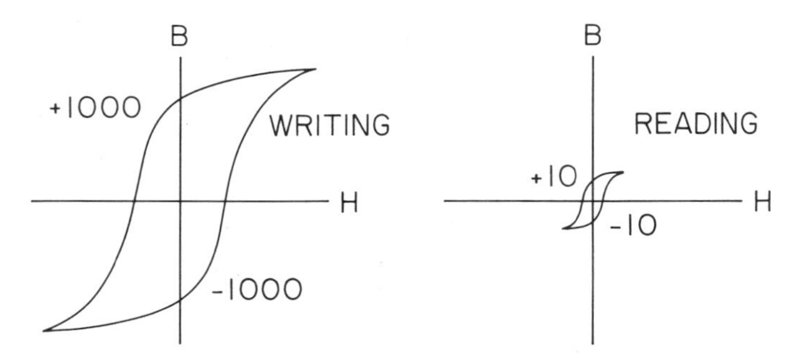

Fig. 4.1. *B–H* loops involved in the writing and reading processes.

fields above the gap. The principal function of a writing-head magnetic material is simply to provide a high enough flux density; other questions such as inductance, efficiency, and energy dissipation, though important to understand, are secondary. When there is too much inductance, more voltage is required; with too low an efficiency, more current is needed and heating is rarely a problem.

For a reproducing head the minor loop of concern is typically of very low flux density, say only 10–20 G. Reproducing heads are more critical because they play an important role in determining the signal-to-noise (SNR) of the system. A low efficiency reduces the output signal voltage and a high dissipation, or minor-loop area, generates high head noise.

In order to facilitate mathematical analysis, it is usually pretended that the reproducing head low-level minor loops have the ellipsoidal shape shown in Figure 4.2. This shape, called a Lissajou's figure, occurs whenever a system responds to a sinusoidal excitation with an in-phase and a 90° out-of-phase delayed response. The permeability is written as a complex quantity,*

$$\mu^* = \mu' - j\mu'' \tag{4.1}$$

where μ^* is the complex permeability, μ' and μ'' are the so-called real and imaginary parts of permeability, and $j = \sqrt{-1}$. This nomenclature need not cause concern. Of course a measured permeability cannot actually be imaginary. Real and imaginary are merely adjectives, albeit unfortunate,

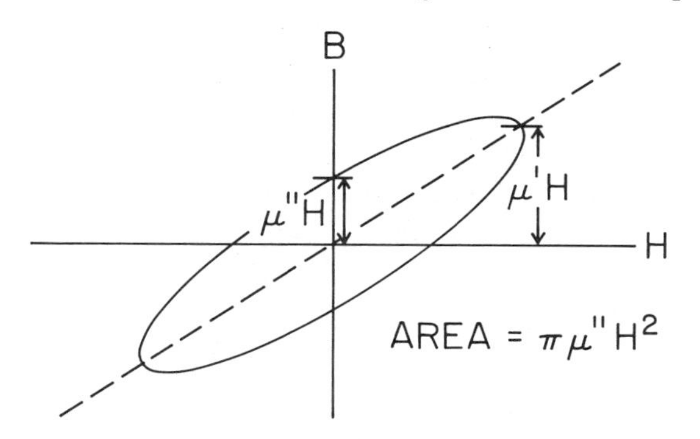

Fig. 4.2. The Lissajou approximation of a low-level B–H loop.

which label the two orthogonal responses. Complex algebra is but a system of performing two orthogonal calculations at once with the j factor keeping them separable.

The complex inductance of the toroid shown in Figure 4.3 is

$$L^* = \frac{0.4\pi 10^{-8} N^2 A \mu^*}{l} \tag{4.2}$$

where L^* is the complex inductance in henries, N is the number of turns, A is the toroidal cross section in square centimeters, l is the average circumference in centimeters, and μ^* is the complex permeability.

The complex impedance is

$$Z^* = j\omega L^* \tag{4.3}$$

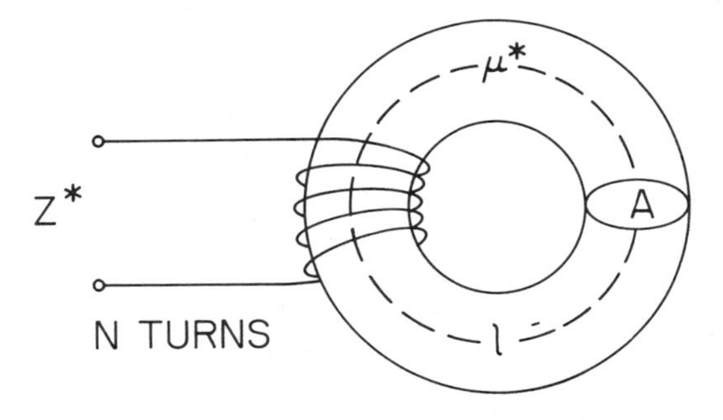

Fig. 4.3. A toroidal sample.

where ω is the angular frequency (rad/sec). The complex inductance may be shown to have two parts, which, omitting the $0.4\pi 10^{-8}$ factor, are

$$Z^* = \frac{j\omega N^2 A\mu'}{l} + \frac{\omega N^2 A\mu''}{l} \tag{4.4}$$

The first part of the equation represents the impedance of a pure inductor. Inductors, like capacitors, merely store energy. The second part, the real part of the impedance, is that of a circuit element like a resistor, which dissipates energy. Clearly, the area of the idealized minor hysteresis loop is proportional to μ'', the imaginary or lossy part of the permeability.

The Nyquist noise theorem asserts that any system that dissipates energy when connected to an electrical power source generates a thermal noise voltage when it is in thermal equilibrium. For a resistor, the noise voltage is given by Johnson's formula,

$$E_n = \sqrt{4kT \cdot \Delta f \cdot R} \tag{4.5}$$

where E_n is the (mean) noise voltage in volts, k is Boltzmann's constant, T is the absolute temperature, Δf is the bandwidth in hertz, and R is the resistance in ohms. The physical reason for this noise voltage is simply that, at any instant in time, unequal numbers of thermally excited electrons are traveling toward each end of the resistor.

The significance of this analysis of minor-loop behavior is in showing that all reproducing heads generate noise that is proportional to the imaginary part of the permeability. Consequently, in reproducing-head magnetic materials, not only is a permeability of high magnitude needed, but the imaginary part must be made as small as possible. The fundamental cause of the dissipation is impediments to the domain-wall motions; the steps discussed in Chapter 2 are employed to reduce the losses. It follows also that, since inductance scales proportional to the physical dimensions, the smaller the head, the less the noise. Naturally, the resistance of the head coil, which is dominant in thin-film heads, also generates Johnson noise. For resistors, the scaling laws are different: the bigger the resistor dimensions, the lower the resistance and the noise.

4.3 Electrical Equivalents of Magnetic Circuits

Because electrical circuits and their analysis are so widely understood, it is, for many, convenient to analyze the flux flow in head structures by using electrical equivalent circuits. Suppose in the toroid shown in the Figure 4.3, the flux density is uniform throughout the cross-sectional area. Then,

$$\mathbf{B} = \mu\mathbf{H} \tag{4.6}$$

$$\phi = \int \mathbf{B} \cdot d\mathbf{A} = BA = \mu HA \tag{4.7}$$

where

$$H = \frac{0.4\pi NI}{l} \tag{4.8}$$

and, therefore,

$$\phi = (0.4\pi NI)\frac{\mu A}{l} \tag{4.9}$$

When this equation is written,

$$Flux = \frac{magnetomotive\ force}{reluctance}$$

and compared to Ohm's law,

$$Current = \frac{electromotive\ force}{resistance}$$

the analogy becomes evident. Magnetic flux is equivalent to electric current; the potential for confusion here is that, by Faraday's law, an electromotive force (emf), or voltage, is generated by a time-changing flux. The magnetomotive force (mmf), $0.4\pi NI$, is analogous to the emf; the likely confusion in this case is that mmf is generated by current. The electrical resistance is equivalent to the magnetic reluctance, $l/A\mu$.

Note that when the coil is driven, as is usually the case in recording, with a high impedance, constant-current source, so that a specific current waveform flows in the coil regardless of the head's impedance, then a constant mmf appears on the coil. This constant mmf is equivalent to a constant voltage generator in the electrical equivalent circuit. In some low-price recorders, constant-voltage write drivers are used in order to save power; in these cases, a constant rate of change of flux occurs and the electrical equivalent is a constant rate of change of current generator.

4.4 Equivalent Circuit of a Head

Figure 4.4 shows a gapped ring head and its equivalent electric circuit. The equivalent circuit is driven by a constant-voltage generator of magnitude, $0.4\pi NI$. One-half of the core reluctance appears on each side of the circuit. Just as the magnetic flux in the head divides into three paths, fringing above and below and going through the gap, the equivalent circuit has three parallel reluctance paths.

In normal heads, the fringing above and below the gap reluctances are very high compared to gap reluctance and can, to first approximation, be ignored. The efficiency of a write head is defined as

$$\text{Efficiency} = \frac{\text{mmf across the gap}}{\text{mmf on the coil}} \tag{4.10}$$

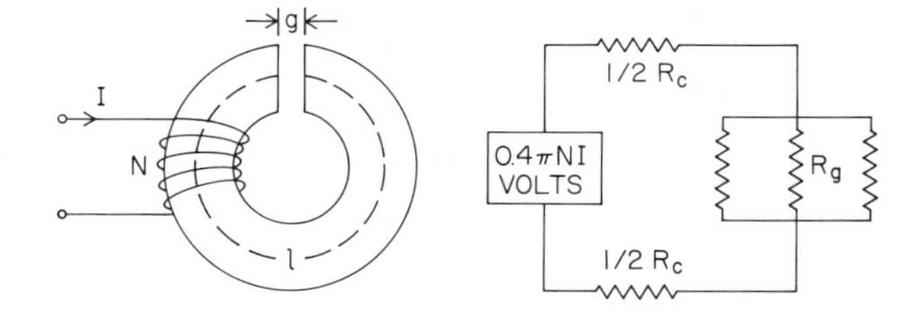

Fig. 4.4. A write head and its equivalent electrical circuit.

and, in order, to calculate its value, the voltage drop across the gap reluctance must be worked out. The current flowing around the circuit is $0.4\pi NI/R_c + R_g$, and the voltage across the gap reluctance is this current times R_g. Thus

$$\text{Efficiency} = \frac{R_g}{R_c + R_g} \tag{4.11}$$

Note that in order for a head to be efficient, the magnetic reluctance of the ring-shaped core must be low in comparison with that of the gap, that is,

$$\frac{l}{A_c\mu^*} \ll \frac{g}{A_g} \tag{4.12}$$

where l is the ring core circumference, A_c is the ring core cross-sectional area, μ^* is the complex permeability, g is the gap length, and A_g is the gap cross-sectional area. In efficient heads, the cross-sectional area of the core is orders of magnitude greater than that of the gap. Moreover, the core circumference is made as small as is possible, this means that the coil windings should "fill the window." A head with excess window area is not as efficient as is possible.

Video heads and computer, rigid disc, heads operate at similar frequencies in the range of 10–20 MHz and have similar efficiencies in the range of 70–80%. As the gap depth, called the throat height in computer peripheral parlance, decreases, for example by wear, the head efficiency increases because the gap reluctance increases. Limiting efficiencies in the range of 90–95% are attained just as the gap wears through. The efficiencies do not reach 100% because the fringing reluctances finally become no longer negligible.

For many reasons, in the analysis of magnetic recording systems, it is necessary to know the deep-gap field,

$$H_0 = \frac{0.4\pi NI \cdot \text{Efficiency}}{g} \tag{4.13}$$

where H_0 is the deep-gap field in oersteds, and g is the gap length in centimeters.

4.5 Head Inductance

The electrical engineer's definition of inductance is

$$E = -L\frac{dI}{dt} \tag{4.14}$$

where E is the voltage in volts, L is the inductance in henries, I is the current in amperes, and t is the time in seconds. By Faraday's law, however,

$$E = -10^{-8}N\frac{d\phi}{dt} = -L\frac{dI}{dt} \tag{4.15}$$

Upon integration in time,

$$10^{-8}N\phi = LI \tag{4.16}$$

and, thus, the inductance,

$$L = 10^{-8}\frac{N\phi}{I} = 10^{-8}\frac{N}{I}\cdot\frac{0.4\pi NI}{R_c + R_g} \tag{4.17}$$

For an efficient head, however, $R_c \ll R_g$, so that the inductance is proportional to N^2/R_g. For a gap of gap length, g, a gap depth, d, and a track width, W, the inductance is

$$L = 10^{-8}0.4\pi N^2\frac{Wd}{g} \tag{4.18}$$

In the manufacture of magnetic heads, the measurement of inductance is the only simple electrical quality control technique possible. Unfortunately, the results of inductive testing can be misleading. Assuming that the heads are efficient, an anomalously high or low inductance could be caused by the wrong number of turns, shorted turns, and/or incorrect gap dimensions, g, W, d. In many heads, it is not possible to count the number of turns (which is, incidentally, always a whole number equal to the number of times the wire goes through the winding window), shorted turns are almost impossible to localize, and often, only the gap length can be measured optically. Moreover, since the efficiency of an efficient head changes only slightly as the gap depth, d, is reduced, it is obvious from Equation 4.13, that the deep-gap field also changes little. Thus, there need be no significant connection between inductance and recording performance.

In some head designs, conductive gap shims, made of copper or silver, are used. At high frequencies, eddy currents flow around these shims in a sense which is opposite to that of the driving current in the coil windings, thereby reducing the flux flowing in the head. The effect in the equivalent electrical circuit is that of putting an opposite polarity voltage generator in series with the gap reluctance. The consequences are two-fold: because less flux flows around the head for a given coil current, the inductance is reduced, and because the effective gap reluctance is increased, the head efficiency is increased. The inductance reduction is principally of concern to designers of the writing and reproducing amplifiers. The efficiency increase can be large for low efficiency heads; for high efficiency heads, the change may be negligibly small.

4.6 Fringing Field Shape

In Figure 4.5 the flux flow, that is the **B** field, is sketched in the region around the gap of a ring head. Notice, initially, that the flux density increases, inside the pole pieces, toward the gap corners. If the gap corners are sharp, then, mathe-

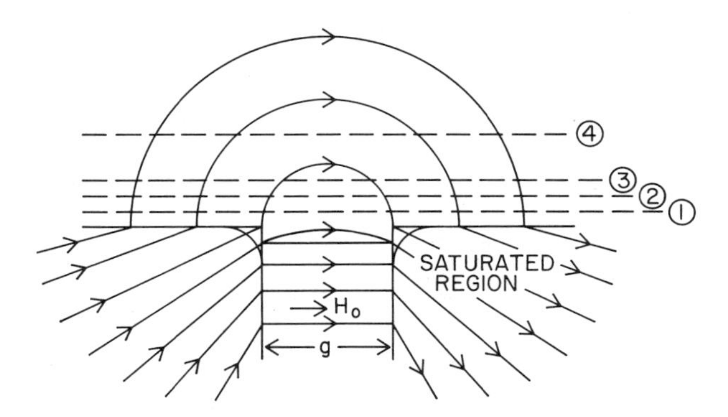

Fig. 4.5. The **B** field adjacent to the gap of a ring head.

matically, the **B** field becomes logarithmically infinite at the corners. In reality, of course, this does not happen; rather, the magnetic material saturates because the material cannot exceed B_s. It may be concluded that even extremely small coil currents cause small regions of the gap corners to be saturated and that, as the current increases, the saturated region grows. When the saturated region size is comparable to about half the gap length, the fringing field above the gap reaches its maximum possible value and further increases of current have little effect except to make the saturated region larger. As an approximation, saturation of the fringing field begins when the deep-gap field reaches about $0.6B_s$. In thin film heads, the saturation flux density is usually reached first at parts of the structure which are remote from the pole tips, and in this case, head saturation occurs without changing the shape of the fringing field.

The fringing field shape above the head may be computed using several more or less difficult techniques, among which are the Schwarz–Christoffel conformal transformation, the Fourier harmonic analysis, and numerical finite element or difference methods. All of the techniques address the central problem of solving Laplace's equation in the region above the head, given the boundary conditions. Usually, the pole pieces are assumed to be made of infinitely permeable ($\mu = \infty$) material, upon which the boundary conditions are

$$B_n(\text{inside}) = B_n(\text{outside}) \tag{4.19}$$

$$H_t(\text{inside}) = H_t(\text{outside}) = 0 \tag{4.20}$$

where B_n is the normal component of the **B** field, and H_t is the tangential component of the **H** field.

In two dimensions, Laplace's equation is

$$\nabla^2 H_x = \nabla^2 H_y = 0 \tag{4.21}$$

where ∇^2 is a linear differential operator formed by taking the divergence of the gradient of a scalar, and H_x and H_y are the x and y components of the magnetic field, **H**. The gradient is, of course, just the slope and is a vector.

Figure 4.6 shows the horizontal, or x, component of the fringing field for the four trajectories, parallel to the top of

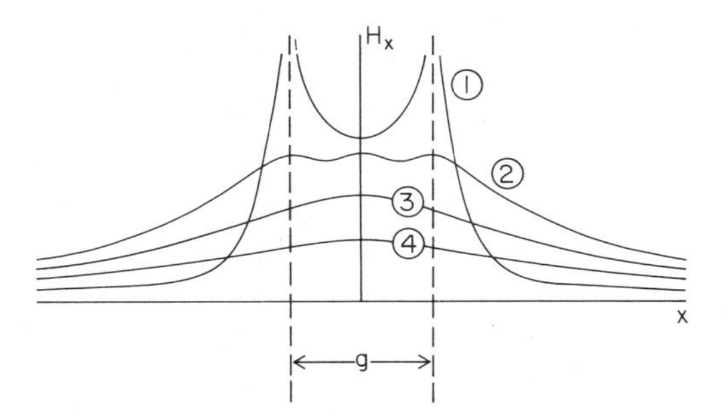

Fig. 4.6. The longitudinal component of the fringing field above the gap of a ring head.

the head, shown in Figure 4.5. For trajectory 1, which is the closest to the head, the field component H_x is sharply peaked above the gap corners. Mathematically, these peaks reach extremely high values, but in physical reality they are limited in magnitude to B_s of the magnetic material used in making the head. Trajectory 2, which is at a distance $y = 0.12g$ above the top of the head, has three peaks: those above the gap corners and that above the center line of the gap. For all trajectories with $y > 0.12g$, only the center-line peak occurs, as is shown for trajectories 3 and 4. For trajectories higher above the head, the peak field decreases but the horizontal extent of the field increases. In fact,

$$\int_{-\infty}^{\infty} H_x \, dx = \text{mmf across the gap} \tag{4.22}$$

and the areas beneath these H_x versus x plots are all equal.

4.7 The Karlquist Approximation

Unfortunately, the many exact derivations of the fringing field above the gap cannot be reduced to simple closed-form mathematical expressions. For most practical purposes, however, Karlquist's simple approximation suffices. Except at points within about one-fifth of the gap length from the gap corners, the differences involved are negligible.

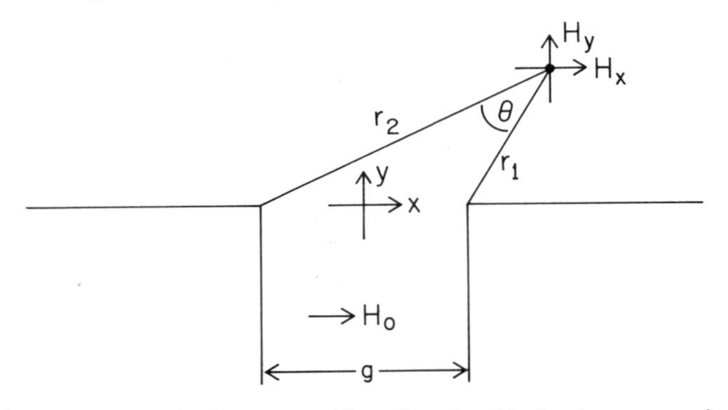

Fig. 4.7. Geometrical construction for the Karlquist approximation.

Figure 4.7 shows the gap region of a head of gap length, g. Any point (x, y), referenced to the origin of the coordinates at the top and on the midplane of the gap, defines two radials, r_1 and r_2, and an included angle θ degrees. For a deep-gap field of H_0 (see Equation 4.13), the horizontal field component is

$$H_x = H_0 \frac{\theta}{\pi} = \frac{H_0}{\pi} \tan^{-1}\left[\frac{yg}{x^2 + y^2 - (g^2/4)}\right] \tag{4.23}$$

and the vertical field component is

$$H_y = \frac{H_0}{\pi} \log_e \frac{r_1}{r_2} = \frac{H_0}{2\pi} \log_e \left(\frac{(x - g/2)^2 + y^2}{(x + g/2)^2 + y^2}\right) \tag{4.24}$$

The great virtues of the Karlquist model are that the geometry of Figure 4.7 can be easily visualized and results obtained quickly by mental estimation.

A particularly useful property of the horizontal field component follows from elementary geometry. Since the angle subtended by any chord of a circle is a constant, it follows that the constant horizontal field contours are nesting circles touching the gap corners as shown in Figure 4.8. It also may be shown that the constant vertical field contours form the set of circles that are orthogonal to those of a constant horizontal field as is indicated in the figure. Note carefully that

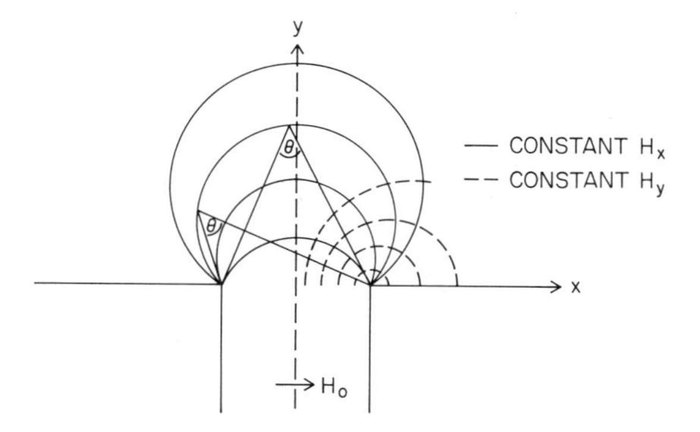

Fig. 4.8. Contours of equal longitudinal and perpendicular field components in the Karlquist approximation.

these depictions of contours of constant-magnitude field component are not field plots.

In the original formulation of the Karlquist model, it was assumed that the magnetic potential falls at a constant rate across the gap, yielding a constant horizontal field, H_x, across the top of the gap. It was later realized that the model also admits two much simpler interpretations, which are in accord with the notion that magnetic fields come from magnetic poles and real currents only.

The first of these interpretations is shown in Figure 4.9. Suppose that the magnetization in the pole pieces is uniform and that sheets of magnetic poles of constant strength but

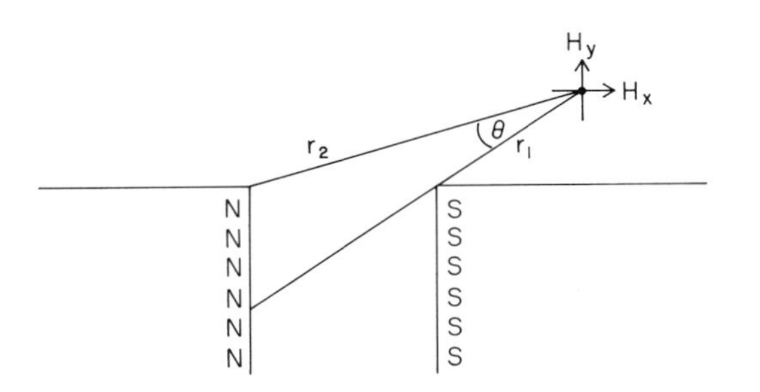

Fig. 4.9. The Karlquist pole model.

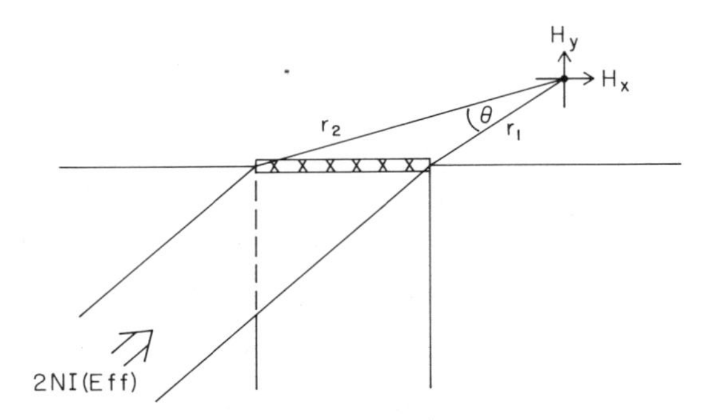

Fig. 4.10. The Karlquist current model.

opposite polarities are generated on the gap faces. These sheets of poles produce exactly the fields given in Equations 4.23 and 4.24, and in fact, only the line-of-sight poles contribute to the field at any point. The line-of-sight poles are those defined by the included angle θ.

The second, and probably more usable, interpretation is the current sheet model of Figure 4.10. Imagine that across the top of the gap there is an infinitely thin conductive shim carrying a uniform current of magnitude $2NI$(Eff), where NI is the ampere turns in the head coil. The extra factor of two arises because Karlquist's formulae give only the fields in the region above the head. This current shim yields exactly the field components of Equations 4.23 and 4.24. This interpretation makes it manifestly obvious that the principal function of the magnetic material in heads is, by transducing the flux, to make it appear that the coil is at the top of the gap. For heads with very small gap lengths, or when interest centers on the long range or far field of a head, the shim's dimensions become negligible; then the field and the field components become just those of the long conductor, discussed in Chapter 1, and

$$H = \frac{(0.2)(2NI)(\mathrm{Eff})}{\sqrt{x^2 + y^2}} \tag{4.25}$$

$$H_x = (0.2)(2NI)(\mathrm{Eff}) \frac{y}{x^2 + y^2} \tag{4.26}$$

and

$$H_y = (0.2)(2NI)(\text{Eff}) \frac{x}{x^2 + y^2} \quad (4.27)$$

The Karlquist current sheet model makes obvious a possible future development in magnetic recording technology which has far-reaching implications. Suppose that instead of constructing a conventional magnetic head, a photo-lithographically defined narrow strip of a superconductive material were used. With the recent developments in superconductive materials, this is becoming increasingly likely, even at room temperature. Current flow in such a strip or shim would generate virtually the same fields as does a conventional head. Accordingly, a superconducting strip head would function as a writing head and, by the Reciprocity Principle discussed in Chapter six, as a reading head. Furthermore, the possibility exists that, by using multiple superconductive strips, heads with novel and useful properties may be produced.

4.8 Field Shape With Conductive Shims

The effect of conductive shims on head efficiency and inductance has already been discussed. Here, the effect of such shims on the fringing field shape (that is, the geometrical form and not the magnitude) is considered.

Suppose that the conductivity of the shim is very high, or that the frequency is very high, so that the skin depth is very small and essentially no flux penetrates the shim. The boundary condition is then, approximately,

$$B_{\text{n}}(\text{inside}) = B_{\text{n}}(\text{outside}) = 0 \quad (4.28)$$

which is rigorously true for a classical superconductor. Note that this condition is precisely the condition that defines a line of force in the usual rules of flux plotting; no **B** field crosses the line of force.

It follows then, that if the conductive shim contour conforms exactly to that of any of the lines of force shown in Figure 4.5, no change in the remaining field shape occurs. As previously discussed, the efficiency and field magnitude may increase, but the geometrical form of the field remains unaltered.

When a gap is filled with a conductive shim, mechanical polishing of the top of the head causes the conductive shim to be flat across the top of the gap. It, therefore, does not conform precisely to the line of force contour, and accordingly, the field shape is altered slightly.

The shape of the fringing field in this case is identical to that of a ring head with no conductive shim, but having zero gap depth, *d*. The exact form of the fringing field is known, in this case, to be that of ellipses with foci on the gap corners; in fact, it differs very little from that of the normal head. Thus, the presence of the conductive shim does little to change the shape of the fringing field.

4.9 Useful Properties of Two-Dimensional Fields

The two interpretations of Karlquist's model discussed above illustrate a general property of all magnetized bodies. For every magnetization pattern with its configuration of magnetic poles, there exists another, different configuration of Amperian, or hypothetical, currents that yields the identical **B** field everywhere. The Amperian current density, **j**, is given by the curl of the magnetization,

$$\mathbf{j} = \nabla \times \mathbf{M} \tag{4.29}$$

The situation is shown in Figure 4.11. In the figure, a permanent magnet, represented by magnetic poles on its ends and Amperian currents on its sides, is compared with a solenoid carrying real currents. All have the same external field, because **B** = **H** in free space. In the permanent magnet, the internal **B** and **H** fields differ, while, of course, the internal **B** and **H** fields of the solenoid are identical. Magnetic field, **H**, comes from magnetic poles and real currents only; Amperian currents produce flux density, **B**, fields.

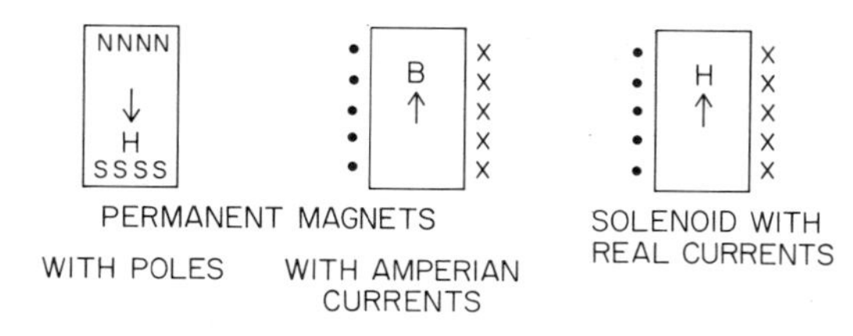

Fig. 4.11. Permanent magnet equivalents and a solenoid compared.

In Chapter 1, it was shown that the magnetic field, **H**, from two-dimensional (2-D) arrays of magnetic poles and real currents are orthogonal. Another, closely related 2-D property is that the 90° clockwise rotation of the magnetization at all points within a magnet results in the exact interchange of the magnetic poles and Amperian currents within that magnet. It follows, then, that the external field at every point rotates by 90° in the opposite direction, that is, counterclockwise. Further, if the magnetization is rotated θ degrees clockwise, the external field rotates by θ degrees counterclockwise.

Any pair of orthogonal field components, such as H_x and H_y, in 2-D space have some interesting and illuminating properties. The Fourier transforms taken along a line parallel to the x axis have the same magnitudes but differ in phase by 90°. Thus the Fourier transforms of Karlquist's Equations 4.23 and 4.24 have precisely this relationship because Karlquist's approximation relates to the sources of the field and not the properties of the field from the approximate source. The H_x field component is an even function, and its Fourier transform contains only even, cosine terms. The H_y field component, on the other hand, is odd and has, therefore, only odd, sine terms. The x and y field components are often called a Hilbert transform pair. Finally, if the magnetization in a magnet is rotated θ degrees at every point, the Fourier transforms of the external field components H_x and H_y remain of the same magnitude, but their phases change by the same θ degrees.

In a later chapter, these arcane, but extremely useful, properties of 2-D fields will be used extensively in the analysis of the reproduce voltage spectra of various recording media–head combinations.

Exercises

1. What is the definition of the efficiency of a writing head?
2. Give an expression for reluctance.
3. Which parameter in an equivalent electrical circuit is analogous to magnetomotive force?
4. Give the mathematical expression for the writing head efficiency.
5. Putting conductors in the gap changes the field shape drastically; true or false?
6. What changes in efficiency occur when conductive gap shims are used?
7. Which regions in a conventional head saturate first?
8. How far above the head does one have to be to get the maximum longitudinal field above the center, or mid-plane, of the gap?
9. Give expressions for the Karlquist fields, H_x and H_y, in Cartesian coordinates.
10. Give an expression for the deep-gap field.
11. The Karlquist field can be thought of as coming from a flat shim, located at the top of the gap, carrying a uniform current. What is the magnitude of the current?
12. Do Amperian current models yield **B** or **H** everywhere?

Further Reading

Karlquist, O. (1954). Calculation of the magnetic field in the ferromagnetic layer of a magnetic drum. *Trans. Royal Inst. Tech.*, in *Introduction to Magnetic Recording* (White, R. M., ed.). IEEE Press, New York, 1985.

Lindholm, Dennis A. (1977). Magnetic fields of finite track width heads. *IEEE Trans. Mag.* **13,** 1460–1462.
Mallinson, John C. (1974). On recording head field theory. *IEEE Trans. Mag.* **10,** 773–775.
Mallinson, John C. (1981). On the properties of two-dimensional dipoles and magnetized bodies. *IEEE Trans. Mag.* **17,** 2453–2460.
Mallinson, John C., and Bertram, H. Neal (1984). On the characteristics of the pole-keeper head fields. *IEEE Trans. Mag.* **20,** 721–723.
Potter, R. I. (1975). Analytical expression for the fringe field of finite pole-tip length recording heads. *IEEE Trans. Mag.* **11,** 80–81.
Westmijze, W. K. (1953). Studies in magnetic recording. *Philips Res. Repts.,* in *Introduction to Magnetic Recording* (White, R. M., ed.). IEEE Press, New York, 1985.

Chapter 5

The Writing, or Recording, Process

5.1 Introduction

In the writing process, the writing-head coil current and fringing field are changed in time as the recording medium is moved past the head. The changing magnetic field magnetizes the tape in a manner that must bear a unique relationship to the writing head coil current. A complete understanding of this process has not been achieved despite many decades of study. There are many reasons for this failure, but there is little doubt that the most profound problem is the almost complete absence of a good 2-D hysteresis model.

In the normal hysteresis loops discussed in Chapter 2, both the ordinate and abscissa fields are measured in the same directions; they are uniaxial loops. In the writing process, a volume element of tape first experiences an increasing positive H_y field component, which rotates to become an H_x field component, then rotates further to become a decreasing negative H_y field component. No model, or useful experimental data, exists today that enables the accurate prediction of, for example, the remanent magnetization vector magnitude and direction after such a field history. This problem is usually avoided in theoretical treatments by considering one field component at a time and adding vectorially the resulting magnetization components. While this procedure bears little relationship to physical reality, it does at least permit some progress to be made.

The second most significant problem is that as the tape becomes magnetized it generates internal, or self-demagnetizing, fields. It follows that the effective writing field is the vector sum of the writing-head field and the self-demagnetizing field. The self-field can only be computed when the magnetization, and its divergence, is known at all points in the tape. Since hysteresis loops are highly nonlinear and multivalued, it follows that numerical iterations are required. The first magnetization is computed using the head field alone, then a second magnetization is worked out using the head field and some fraction, called the convergence factor, of the first self-field, and so on until the computation converges on a stable solution. Such iterative computations are often called self-consistent models.

In a complete computation, consideration of the effect of the proximity of the highly permeable pole pieces, of the writing head, on the self-demagnetizing fields, is also necessary. In order to meet the boundary conditions given in Chapter 4 on the highly permeable pole pieces, it is as though a mirror image, with opposite polarity, of the tape magnetization is formed in the pole pieces. This has the effect of reducing both the x and y components of the self-demagnetizing field. In a later chapter it will be shown, however, that in most practical recorders, the head–medium interface and recorded wavelengths or digital bit densities are arranged so that the effects of self-demagnetizing fields are, in fact, rather small. Self-consistent computations are not discussed further in this book. In practice, they are so complicated that very little physical understanding of the writing process can be derived from them. In this chapter, only the results of extremely simple writing models are discussed because it is believed they provide significant insights into the writing process.

5.2 The Bauer–Mee Bubble Model

Suppose that the writing process is simplified in the following ways: only longitudinal or horizontal fields are consid-

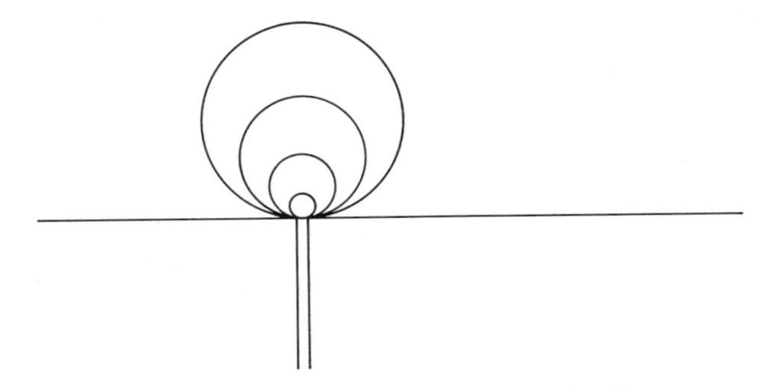

Fig. 5.1. Contours of constant longitudinal field component for a zero-gap ring head.

ered; no self-demagnetizing fields are taken into account; the gap length of the writing head is almost zero; the head-to-medium spacing is zero; the medium's coating thickness is thicker than the depth of recording; and finally, the remanent magnetization loop is ideally square. The (almost) zero gap-length approximation makes the constant horizontal field contours form the nesting circles shown in Figure 5.1. The idealized square M_r–H loop shown in Figure 5.2 has a remanent coercive force, ${}_rH_c$. It follows, from Equation 4.26, that the diameter of the contour of horizontal field equal to ${}_rH_c$ is directly proportional to the writing-head current, I.

Consider the case in which the writing head current, I, is a

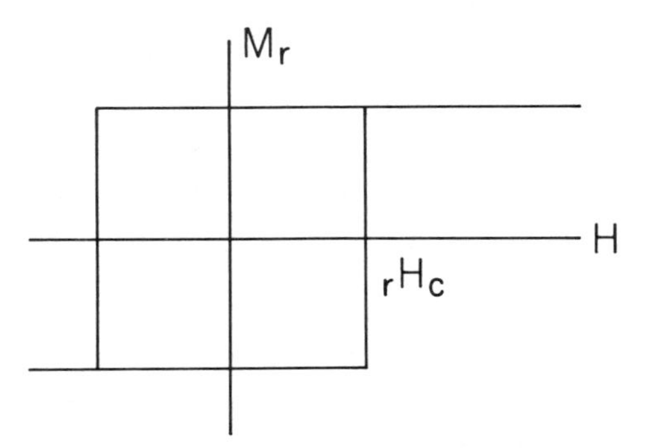

Fig. 5.2. An idealized rectangular remanent magnetization loop.

perfect sinusoid

$$I = I_0 \sin \omega t = I_0 \sin kx \tag{5.1}$$

where I_0 is the peak amplitude in amperes, ω is the angular frequency in rad/sec, t is the time in seconds, k is the wave number $(2\pi/\lambda)$ in cm^{-1}, λ is the wavelength in centimeters, and x is the longitudinal distance in centimeters. Note carefully the transformation in this equation from the temporal, or time, domain to the spatial, or distance, domain. When the recording medium moves past the head at a relative velocity V, cm/sec,

$$V = \omega/k \tag{5.2}$$

When the head–medium spacing is zero, all the material written in the contour of field equal to ${}_rH_c$ is magnetized, or switched, to the full value of the remanence having the same polarity as the write current. For a previously erased tape, a pattern of circles is formed for long wavelength recording, as is shown in Figure 5.3. Note that the contour joining the vertical, or y axis, extent of the circles is given simply by

$$Y = A|\sin kx| \tag{5.3}$$

where A is proportional to the write current and $|\ |$ indicates magnitude only. The appearance of the sequential circles is suggestive of a succession of soap bubbles, hence the appellation "bubble model."

It is of particular importance to note that the remanent flux,

$$\phi_r(x) = B \sin kx \tag{5.4}$$

where B is a constant of proportionality, is precisely propor-

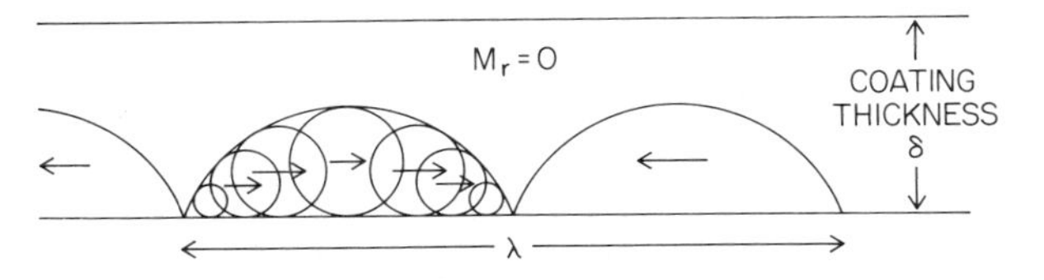

Fig. 5.3. The long-wavelength writing process.

tional to the write-current sinusoid, providing the bubbles do not penetrate through the back of the coating. It is remarkable indeed that, given the extreme nonlinearity of the square M_r–H loop, linear recording has been achieved. Moreover, if the input write current consisted of a number of superimposed sinusoids, the remanent flux would have contained all those sinusoids in proportion and no others.

The mathematical definition of a linear system is

$$L[ax(t) + by(t)] = aL[x(t)] + bL[y(t)] \tag{5.5}$$

where L is a linear transform, a and b are arbitrary scale factors, and x and y are arbitrary signals of time, t. Observe that in the equation the output, or right-hand side, has no terms involving higher powers (x^2, y^2, etc.) or cross products (xy, x^2y^3, etc.). The Bauer–Mee bubble model is, therefore, precisely a linear system.

When either the head–medium spacing is not zero or the recording bubble penetrates through the coating thickness, it may be shown that a form of nonlinearity, called odd-harmonic distortion, arises. However, since the phases of the odd-harmonic distortion caused by these two conditions are, in fact, 180° apart, it turns out that compensation of the odd-harmonic distortion is possible. For every head-medium spacing, there exists a write-current amplitude that minimizes the odd-harmonic distortion. Similarly, in the ac-biased audio recorders discussed in Chapter 8, the ac-bias current amplitude may be adjusted to minimize the odd-harmonic distortion.

When the bubble model is applied to short-wavelength recording, the bubbles begin to overlap and the situation becomes more complicated. The result, for wavelengths short in comparison with the bubble diameter, is shown in Figure 5.4. It is useful because it introduces the idea that the

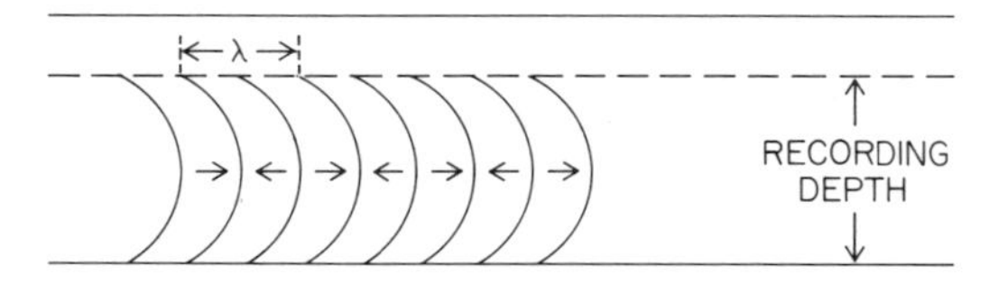

Fig. 5.4. The short-wavelength writing process.

recorded magnetization need not be in phase at all depths in the coating. For a tape moving to the right, the remanence deep in the tape is recorded further to the right, which corresponds to earlier in time, or leading in phase, than that on the surface. The phase angle, in radians, is

$$\Delta\theta = 2\pi \frac{\Delta x}{\lambda} \tag{5.6}$$

where Δx is the x axis displacement.

When $x = \lambda/2$, the phase angle is π radians or 180°, and the remanent magnetization in such a layer subtracts from the remanent flux of the recording. The possibility of complete cancellations at particular combinations of write current and short wavelengths is obvious; these are called recording, or writing, nulls.

5.3 Computer Modeling with the Priesach Function

Useful further insights into the writing process can be gained by modest numerical or computer modeling of the process where actual M_r–H loops are used and the gap length is nonzero.

While it is clear that the Karlquist expressions cover the nonzero, gap-length condition, the use of the actual family of major and minor M_r–H loops is more complicated. One alternative is to devise a set of rules, preferably using simple closed-form expressions, for navigating within the M_r–H plane. A more physical approach is to use the Priesach function method.

The Priesach method uses a density function, called the Priesach function, which is plotted in a plane defined by positive and negative magnetic fields, as is depicted in Figure 5.5. The Priesach function may be related to the switching fields of individual magnetic particles and the magnetic fields, called interaction fields, between particles. In this discussion, however, the Priesach function is regarded, phenomenologically, simply as a function of positive and nega-

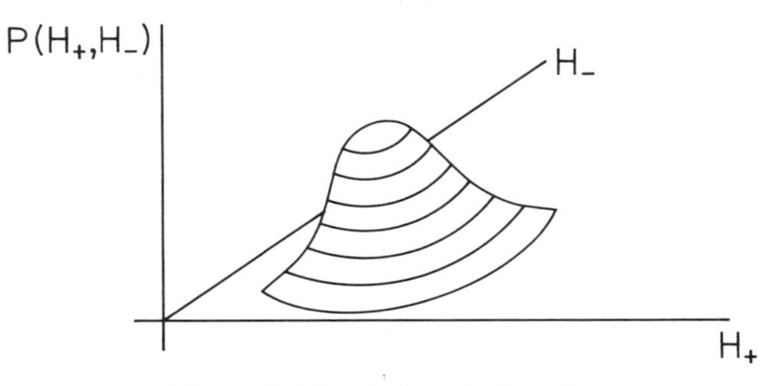

Fig. 5.5. The Priesach function.

tive field extrema, $P(H_+, H_-)$. It may be determined experimentally in a VSM by applying a systematic sequence of H_+ and H_- fields. A typical element of the Priesach function is obtained by measuring the remanence after the sequences H_{1+}, H_{2-} and $(H_1 + dH)_+$, $(H_2 + dH)_-$ fields. It may, accordingly, be regarded as a differential representation of the M_r–H loops.

Suppose that the remanent magnetization obtained by a sequence of field extrema, H_{1-}, H_{2+}, H_{3+}, and H_{4-}, is required. Lines, parallel to the appropriate positive or negative field axis, are drawn; these lines divide the Priesach function into two regions separated by a line called the terminator, as is indicated in Figure 5.6. The action of the terminator is analogous to that of a cookie cutter. The volume, obtained by integration, of the upper region is proportional to positive remanence and the lower region is proportional to negative remanence. The total remanent magnetization obtained by that field sequence is

$$M_r = \int_{\text{upper region}} P\, dH_+\, dH_- - \int_{\text{lower region}} P\, dH_+\, dH_- \quad (5.7)$$

It is obvious that when many field extrema symmetric about zero field as applied, as is the case in the ac demagnetization process, that the volumes above and below the terminator are equal and a remanent magnetization of zero results. When there is also a small dc field, however, as is the case in the process of anhysteresis, a nonzero difference in volumes results; it is proportional to the anhysteretic remanence.

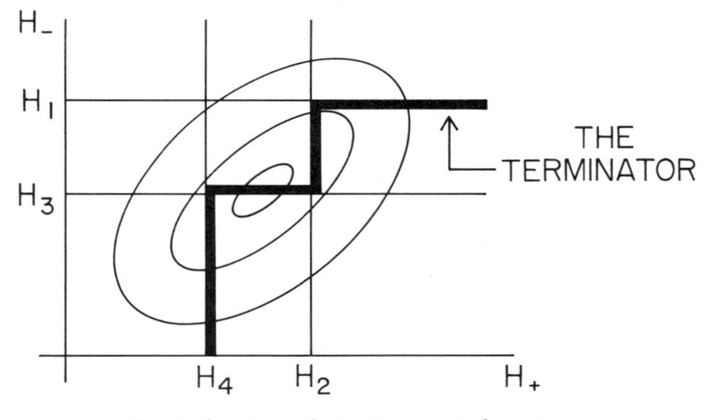

Fig. 5.6. Use of the Priesach function.

The particular utility of the Priesach function is that, since each volume element of the recording medium experiences a different sequence of fields as it passes the writing head, the remanence of each element may be computed easily. Figure 5.7 shows the field histories experienced by two adjacent volume elements of a tape during this trajectory parallel to the top of a ring head for a sinusoidal writing current. The field history is given by multiplying the input sinusoid by the horizontal field function and results in the sequence of field

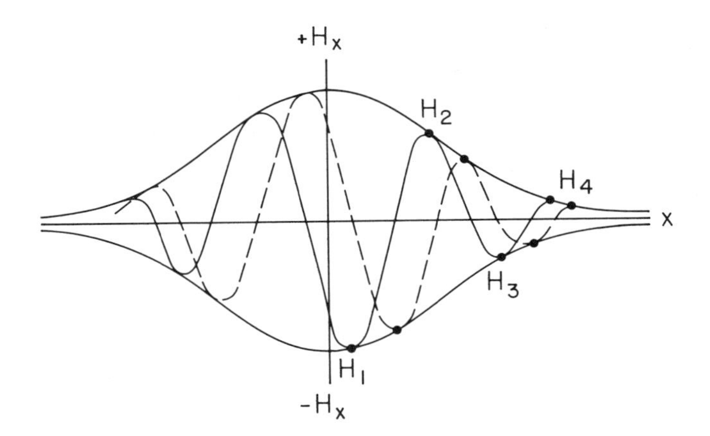

Fig. 5.7. Sequences of field extrema of alternating polarity and decreasing magnitude which determine the written remanent magnetizations.

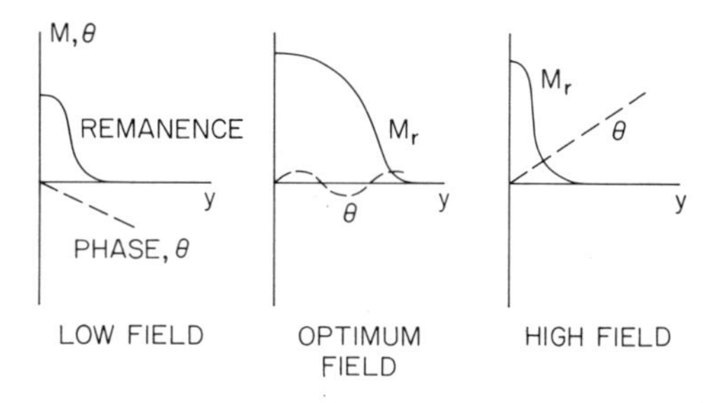

Fig. 5.8. Plots of the amplitude and phase angle of the remanent magnetization versus depth in the tape coating for different writing-field strengths.

extrema of alternating polarity shown. The sequence of extrema for the two points are slightly different, reflecting the different apparent phase of the input sinusoid. Moreover, the field extrema are different for points at differing depths in the medium. The sequence of field extrema for each volume can be mapped directly into the Priesach plane as shown in Figure 5.6.

The results of such numerical modeling appear in Figure 5.8, which shows the magnitude and phase of the remanent magnetization versus depth for three magnitudes of the input sinusoid. Figure 5.9 shows the contours of constant longitudinal components of the write field corresponding to the three levels of write current. Contours of $H_x = 200$ Oe and 500 Oe are chosen because they approximate the negative and positive 90% of maximum remanence fields for a standard γ-Fe_2O_3 medium with a coercivity of ${}_rH_c$ equal to 300 Oe. The difference, 300 Oe, is called the range of switching

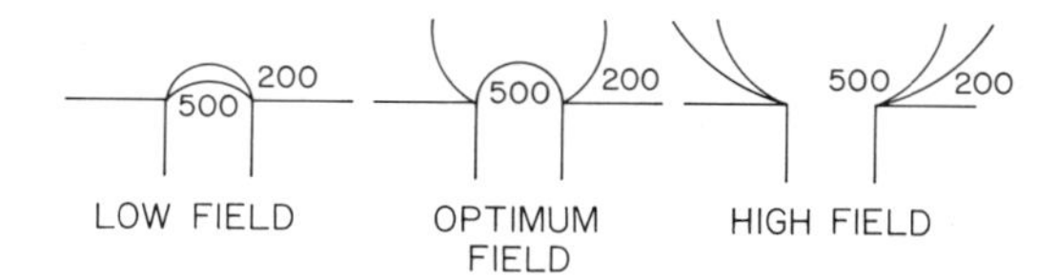

Fig. 5.9. The writing-zone geometry for different writing-field strengths.

fields, ΔH, of the medium; it is usually proportional to the coercivity.

At the low drive, or writing current, level, the bubble-like contours and the magnetization do not penetrate deeply into the tape and the recording phase fronts slope to the left; that is, the recording lags in phase. At high drive levels, the phase fronts slope to the right causing the phase to lead with increasing depth, and additionally, the remanence decreases appreciably deeper in the coating. The remanence decrease is attributable to the increasing distance between the 200 and 500 Oe contours. The greater the distance, the lower is the writing-head field gradient, dH_x/dx, the more decreasing-amplitude field extrema occur, and the more the situation resembles ac demagnetization. This phenomenon has, confusingly enough, sometimes been called recording demagnetization; low write-field resolution seems to be a better name.

Note particularly the conditions that prevail when the write current is neither too low nor too high but is optimum. Now the magnetization is high deep into the medium and the phase angle is almost zero. This zero phase error occurs because the field contours are, more or less, perpendicular or normal to the head plane and the remanent magnetization is written at the same longitudinal position at all depths. The optimum condition, which results in the maximum possible reproduce output voltage at short wavelengths, occurs when the higher field contour forms almost a semicircular bubble above the gap. In the optimum condition, the distance between the contours, which governs the write-field resolution, is almost independent of the gap length. The optimum write current is such that the deep-gap field is

$$H_0(\text{optimum}) = 3{}_rH_c = \frac{0.4\pi NI \cdot \text{Eff}}{g} \tag{5.8}$$

Thus for standard γ-Fe_2O_3, the optimum short wavelength deep-gap field is about 1000 G, for cobalt-surface-modified-γ-Fe_2O_3 it is 2000 G, and for iron particles, some 5000 G is required. Given that the B_s of ferrite is 5000 G, it may be concluded conversely that the maximum possible coercivity is 1600 Oe, and that further, in order to avoid pole-tip saturation effects, (0.6)1600 = 1000 Oe is the maximum desirable coercivity in a 100% efficient head. These estimates are in good accord with practice.

5.4 The Williams–Comstock Model

In many applications, particularly in the analysis of computer peripheral flexible and rigid discs, where the coating is extremely thin in comparison with that in tape, the Williams–Comstock model is applicable. In this model, the following simplifications are used: only longitudinal writing-head fields are considered, the magnetization is supposed to be uniform throughout the coating and to be longitudinally oriented only, and the longitudinal demagnetizing field is treated in an approximate manner. The model will be explained in considerable mathematical detail because it is so widely used and discussed.

The first basic idea in this model is that a specific form,

$$M_x(x) = \frac{2M_r}{\pi}\tan^{-1}\left(\frac{x}{f}\right) \tag{5.9}$$

is assumed for the variation of the recorded magnetization in the x direction in response to a step function change in the write current. The arctangent transition has the advantage that it can be specified by only one factor, f, called the arctangent parameter. As is shown in Figure 5.10, the maximum slope of the arctangent function occurs as it passes through zero and is proportional to f^{-1}. There appears to be little physical justification for the arctangent transition; as will be shown here and in Chapter 10, its appeal is principally that of mathematical convenience.

The second basic idea in the model is that only the slope, dM_x/dx, as it passes through zero, will actually be computed according to

$$\frac{dM_x}{dx} = \chi\left[\frac{dH_h}{dx} + \frac{dH_d}{dx}\right] \tag{5.10}$$

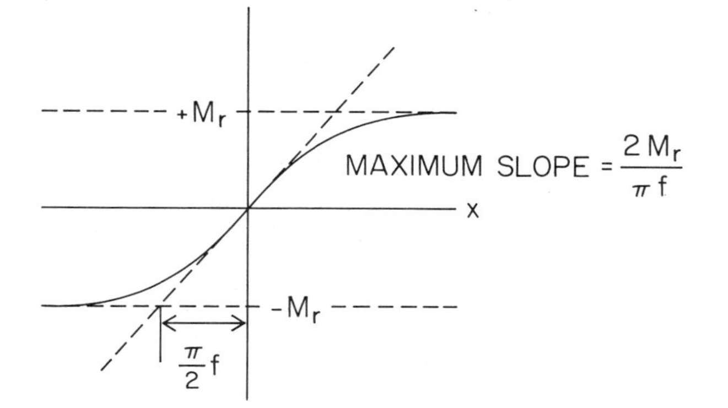

Fig. 5.10. The arctangent magnetization reversal.

where χ is the remanent susceptibility (measured at ${}_rH_c$ on the major loop), H_h is the head field, and H_d is the (negative) demagnetizing field.

The head-field gradient, dH_h/dx, of course, depends on the write current and the position, (x, y), in the medium. In the Williams–Comstock model, the following procedure is used to determine the maximum head-field gradient. Consider Karlquist's equation for a zero-gap head,

$$H_h(x) = 0.4I \frac{y}{x^2 + y^2} \tag{5.11}$$

Set y to correspond to the midplane of the medium; set $H_h = {}_rH_c$; calculate dH_h/dx; calculate d^2H_h/dx^2 and set it equal to zero, yielding a specific value of x, and finally, substitute the value of x into dH_h/dx. This procedure finds the x coordinate, $(x = y)$, where the head field has simultaneously a value equal to ${}_rH_c$ and the maximum gradient, as is shown in Figure 5.11. It turns out that the maximum gradient is

$$\left(\frac{dH_h}{dx}\right)_{\max} = \frac{-{}_rH_c}{y} \tag{5.12}$$

and that this occurs at the optimum write current,

$$(I)_{\text{opt}} = 5y{}_rH_c \tag{5.13}$$

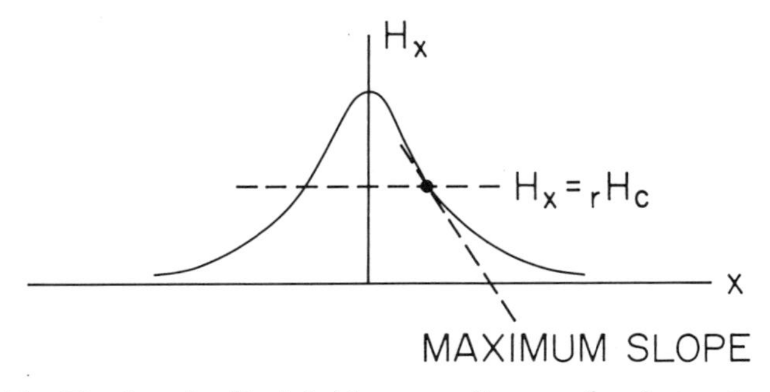

Fig. 5.11. The longitudinal field versus distance for the optimum writing condition.

These equations are of great interest because they show clearly several universal truths about the writing process. First, note that the higher the remanent coercivity, ${}_rH_c$, the greater becomes the magnitude of the maximum field gradient. The length of the recording zone is, of course, the gradient multiplied by the range of fields, ΔH, over which the medium switches magnetization. In most media, ΔH is proportional to the coercivity and it follows that the record zone length, or spatial resolution of the writing process, is almost independent of the coercivity. Usually, as the coercive force is raised, the optimum write current increases proportionally, as is shown by Equation 5.13, and therefore the geometry of the recording process remains unchanged. Another point, to be noted from Equation 5.12, is that the smaller the value of y, or the thinner the recording medium, the higher the maximum head-field gradient. The writing-process spatial resolution increases monotonically as the head–medium spacing and the coating thickness are reduced.

The demagnetizing field gradient, dH_d/dx, is determined by the following procedure: from the arctangent transition, obtain the pole density, $\rho = -\nabla \cdot \mathbf{M} = -dM_x/dx$; integrate to find the demagnetizing field as a function of x; differentiate to get dH_d/dx as a function of x; and finally, set x equal to zero. As is shown in Figure 5.12, for a thin medium, the result is that the maximum demagnetizing field gradient is,

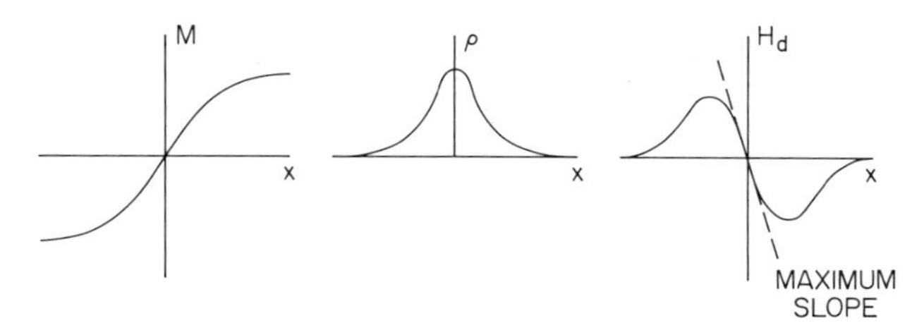

Fig. 5.12. Plots of the arctangent magnetization, pole density, and demagnetizing field versus distance.

$$\left(\frac{dH_d}{dx}\right)_{\max} = -\frac{4M_r\,\delta}{f^2} \qquad (5.14)$$

where δ is the (thin) coating thickness.

When the susceptibility of the recording medium is very high, then to a good approximation, the writing head and demagnetizing field gradients must be equal and opposite. In this case, a further simple analytical result follows. The arctangent parameter is

$$f = 2\sqrt{\frac{M_r}{{}_rH_c}\,\delta\left(d + \frac{\delta}{2}\right)} \qquad (5.15)$$

where d is the head–medium spacing.

In a standard γ-Fe_2O_3 disc drive, where ${}_rH_c = 325$ Oe and $4\pi M_r = 800$ G, the head–disc spacing is small (about 0.3 μm) in comparison with the coating thickness (about 0.8 μm), and the arctangent transition parameter, f, is approximately equal to the coating thickness, δ.

The Williams–Comstock model is useful, not only because it permits simple mathematical analysis, but also because experimentally recording data, specifically the reproduce voltage spectra, may be understood in a particularly simple manner. This topic is treated in Chapter 10.

Exercises

1. Which of the following simplifications are used in the Bauer–Mee bubble model: zero switching-field distri-

bution, zero gap length, zero head–tape spacing, and coating thicker than the maximum record depth?

2. Why is the long wavelength recording in the Bauer–Mee bubble model almost a linear process?
3. What fault in the recording process can give rise to recording nulls?
4. Give an expression for the recording zone length in terms of the head-field gradient and the range of switching fields, ΔH, of the medium.
5. Give a mathematical definition of a linear system.
6. Why is the optimum write-process geometry almost independent of the coercive force of the medium?
7. What is the maximum longitudinal field above a 100 microinch gap, when the point of measurement is 50 microinches above the gap and the deep-gap field is 750 Oe?
8. What is the optimum short-wavelength, deep-gap field for an iron particle tape with ${}_{r}H_{c} = 1450$ Oe?
9. How does the write head, longitudinal field maximum gradient vary with the distance above the top plane of the head?
10. For thin media of high susceptibility, give an expression for the arctangent parameter, f, according to the Williams–Comstock model.
11. What is the most important missing thing which prevents a definitive analysis of the writing process?
12. Give an expression that represents the fundamental idea used in the Williams–Comstock model.

Further Reading

Bauer, B. B., and Mee, C. D. (1961). A new model for magnetic recording. *IRE Trans. Audio* 61–68.
Schwantke, G. (1961). The magnetic recording process in terms of the Priesach Representation. *J. Audio Eng. Soc.* **9**, 37–47.
Williams, M. L., and Comstock, R. L. (1971). An analytical model of the write process in digital magnetic recording. *17th Annual AIP Conf. Proc.* 738–742.

Chapter 6

Reading, or Reproducing, Processes

6.1 Introduction

In complete contrast to the writing process, the reading, or reproducing, processes are almost completely understood. This is principally because the flux density levels are so low in a reproduce head (only about 10 G) that the process may be considered to be linear. Analyses of considerable elegance and utility have been developed to handle a wide variety of reproduce-process problems. In this chapter, the fringing fields from prerecorded tapes, the reproduce head flux, the reproduce head efficiency, the reciprocity theorem, and several types of reproduce head gap-loss effects are discussed.

6.2 Tape Fringing Fields

Suppose that the recording medium has been written with a perfect, undistorted sine wave of longitudinal magnetization, which has the same magnitude and phase angle throughout the coating depth. If $M = M_0 \sin kx$, then the pole density is

$$\rho = -\nabla \cdot \mathbf{M} = -kM_0 \cos kx \tag{6.1}$$

as is indicted Figure 6.1. The magnetic field, **H**, obtained by integration over these poles, is shown schematically in Fig-

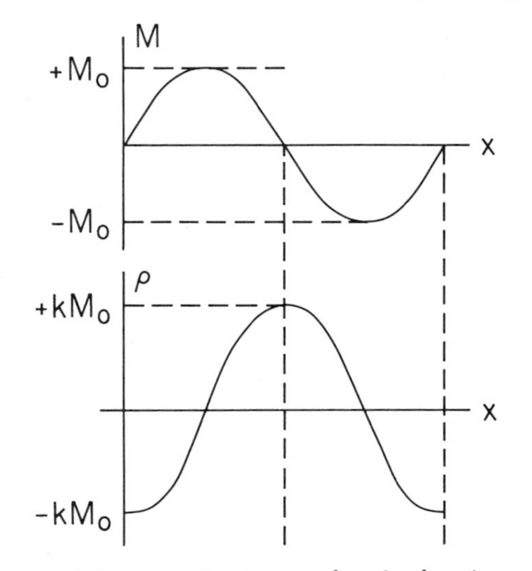

Fig. 6.1. Sinusoidal magnetization and pole density versus distance.

ure 6.2. Analytically, the horizontal component of the fringing field is, at points both above and below the tape,

$$H_x(x, y) = -2\pi M_0(1 - e^{-k\delta})e^{-ky} \sin kx \qquad (6.2)$$

where δ is the coating thickness and y is the magnitude of the distance from the tape surfaces. The vertical component of the fringing field is

$$H_y(x, y) = \pm 2\pi M_0(1 - e^{-k\delta})e^{-ky} \cos kx \qquad (6.3)$$

where the + sign holds at points below the tape and the − sign holds at points above the tape.

Note that, as discussed in Chapter 4, the two field components are in phase quadrature; H_x and H_y are a Hilbert transform pair. Further, the sense of the 90° phase shifts are opposite, above and below the tape. For a trajectory to the right, above the tape and parallel to it, the fringing field is a constant amplitude vector that rotates clockwise; below the tape, the rotation is counterclockwise.

The factor $\exp(-ky)$, which appears in Equations 6.2 and 6.3, is of central and critical importance in magnetic record-

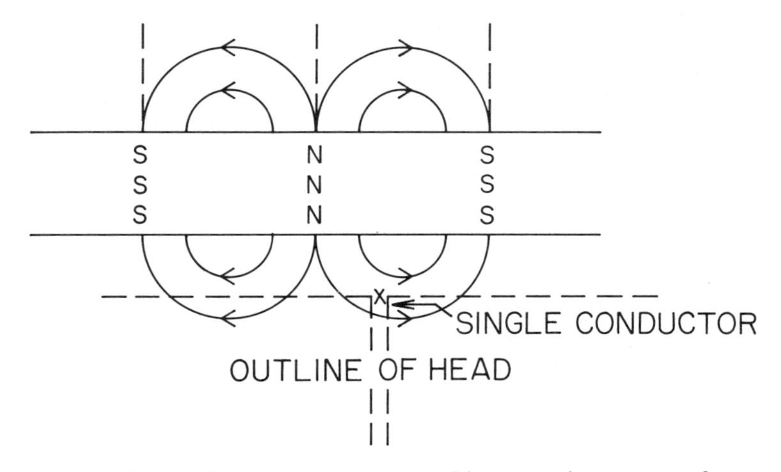

Fig. 6.2. The reading process visualized by the substitution of the read head by its current equivalent.

ing. When the spacing from the tape surface is one wavelength

$$e^{-ky} = e^{-2\pi} = \frac{1}{500} \tag{6.4}$$

The fringing field magnitude, at a distance equal to one wavelength from the tape surface, is only 0.2% of its value on the surface. This rapid attenuation of the fringing field with distance above and below the medium is an inevitable consequence of Laplace's equation in two dimensions and effects the design of magnetic recording systems more than any other single factor.

When the magnetization in the tape is rotated everywhere by an angle θ counterclockwise, the external field at all points in space is rotated by an equal angle θ clockwise. Thus, a rotation of 90° counterclockwise, which makes the magnetization become vertically oriented, causes the fringing fields above and below the tape to lag and lead, respectively, by a 90° phase angle.

6.3 The Reproduce-Head Flux and Voltage

Suppose that an (almost) zero gap-length reproduce head is placed against the tape shown in Figure 6.2. The simplest way to calculate the head flux is to imagine that this read head is replaced, or substituted, by one of the Karlquist equivalent models discussed in Chapter 4. Figure 6.2 shows the substitution of the actual head by the single conductor, which gives the identical head field. The reproduce-head flux can now be obtained by finding, by integration, the tape fringing field that threads, that is, loops around, the conductor. Thus,

$$\phi = W \int_{-d}^{-\infty} H_x \, dy = -4\pi M_0 W \left(\frac{1 - e^{-k\delta}}{k} \right) e^{-kd} \sin kx \tag{6.5}$$

where W is the track width in centimeters, M_0 is the peak sinusoidal remanence in gauss, k is the wavenumber in cm^{-1}, δ is the coating thickness in centimeters, and d is the head-medium spacing in centimeters.

The reproduce-head output voltage, E, is given by the application of Faraday's law,

$$E = -10^{-8} N \frac{d\phi}{dt} = -10^{-8} NV \frac{d\phi}{dx} \tag{6.6}$$

where N is the number of turns on the head and V is the head–medium relative velocity in cm/sec. The result, for the output voltage spectrum is

$$E = 10^{-8} VN(\text{Eff})(4\pi M_0 W)(1 - e^{-k\delta})(e^{-kd})(\cos kx) \tag{6.7}$$

Each of the bracketed terms is discussed, in turn, below. The first such term is the reproduce-head efficiency; it is the fraction of the tape flux entering the head which actually threads the coil.

The second bracketed term is called the tape's short-cir-

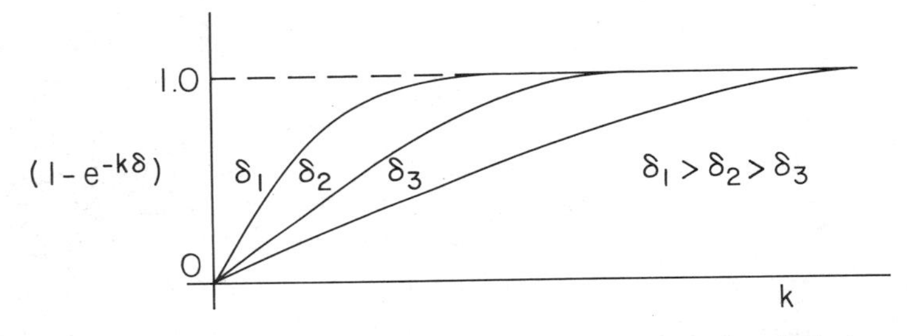

Fig. 6.3. The thickness loss versus wave number for differing thickness.

cuit flux. It is the peak value of the sinusoidal remanent flux "flowing" in the tape per unit of coating thickness.

The third term is, somewhat misleadingly, called the thickness loss. Its behavior is shown in Figure 6.3. Note that at long wavelengths, or small values of the wave number, the thickness-loss term is proportional to $k\delta$. At short wavelengths, however, the thickness-loss term becomes equal to unity regardless of the coating thickness. This is because layers deep in the tape do not contribute appreciably to either the reproduce-head flux or the output-signal voltage. By setting $(1 - e^{-k\delta})$ equal to 0.8 and 0.9, in turn, it can be shown that 80% and 90% of the output signal is generated in shallow layers on the tape surface with depths of 0.37λ and 0.26λ, respectively.

The fourth bracketed term is called the spacing loss and is a direct consequence of the exponential fall off of the tape fringing field discussed previously. In electrical engineering, it is common to use the logarithm of power ratios to compare quantities. The dimensionless unit is called the bel, because its first application was to make measurements of human hearing acuity; ten times the logarithm of a power ratio is called a decibel, written dB. Thus

$$10 \log_{10} P_1/P_2 = 20 \log_{10} E_1/E_2 = \text{dB} \tag{6.8}$$

where P_1 and P_2 are the two powers to be compared and E_1 and E_2 are the two corresponding voltages. The voltages are to be imagined as being applied across a hypothetical one-ohm resistive load. Expressing the spacing loss in decibels yields,

$$20 \log_{10} (e^{-kd}) = -54.6\ d/\lambda \text{ dB} \qquad (6.9)$$

For every wavelength of spacing, the output power and voltage drop 54.6 dB.

As was remarked on earlier in the chapter, the phenomenon of spacing loss is the single most important and critical factor in magnetic recording technology. At short wavelengths, or high digital bit densities, it is crucial that the read–head to medium spacing be made, and kept, as small as possible in order to minimize the spacing loss. Because it is not possible to focus magnetic fields in the same sense that is possible with electromagnetic radiation (light), the spacing loss appears to be inevitable and unavoidable. With nonplanar recording media, such as exist in the obsolete wire recorders, the spacing losses are even more drastic; in three-dimensional solutions of Laplace's equation, Bessel's functions replace the exponentials and they are even more highly dependent on the spacing, d, between the reading head and the media.

The final bracketed term, $\cos kx$, shows that the output signal of a longitudinally magnetized tape is 90° out of phase, lagging the sinusoidal input signal. Whereas the read-head flux is 180° out of phase, the output voltage is 90° out of phase; the difference is due to the 90° phase shift that occurs during the differentiation of the head flux. This 90° phase error has a very important bearing on the output-signal processing operations that occur, of necessity, in all recorders. The reproduce-signal processing must include a 90° phase shifting operation in order to correct this 90° phase error. In analog recorders, this correction is usually made by an integrator (−90°), whereas in digital recorders it is more common to use a differentiator (+90°).

In Figure 6.4, the effects of the thickness and spacing losses on the output-voltage spectrum are shown. The spectrum starts, at long wavelengths, by rising in proportion to the frequency; this is equivalent to +6 dB per octave or 20 dB per decade of frequency. The characteristic then levels

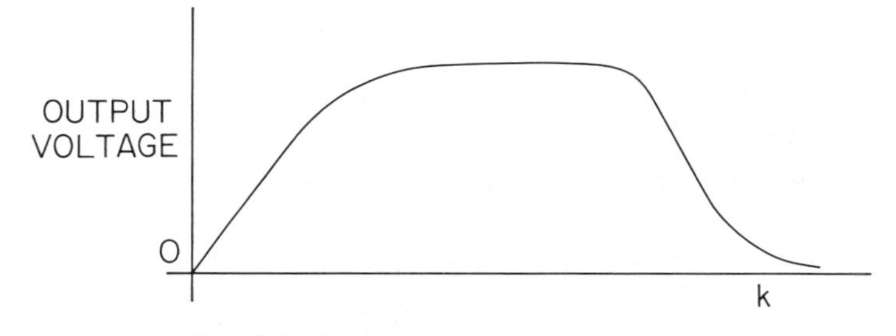

Fig. 6.4. The output-signal spectrum.

off due to the onset of the thickness-loss term and then falls, more or less precipitously, under the influence of the spacing loss.

Figure 6.5 shows the general properties of the flux-density field, **B**, in and around the reproduce head. For the sake of clarity, the **B** field inside the tape is not shown; instead only the magnetic poles are indicated. Note that at points distant from the head, the flux flows equally above and below the tape. Over the head pole pieces, however, most of the flux is "sucked into" the read head. For a head of permeability, μ, the flux density is increased by the factor $2\mu/\mu + 1$, which for any value of $\mu > 10$ is essentially equal to the extra factor

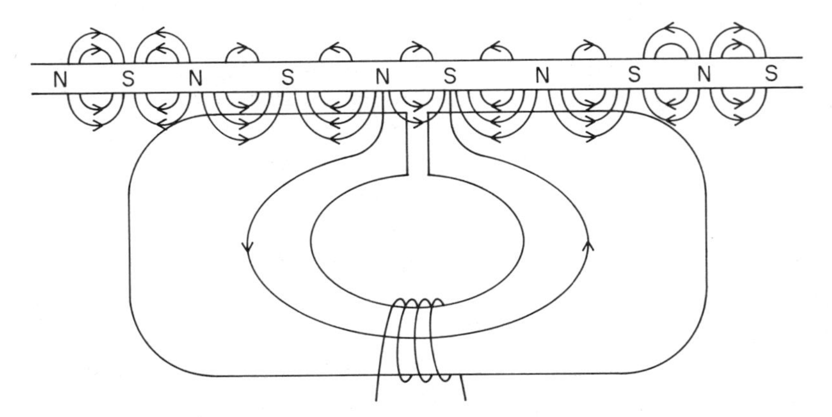

Fig. 6.5. The **B** field outside a tape and within a reading head of less than 100% efficiency.

of 2 which appeared previously in Chapter 4, for the Karlquist equivalent conductor model.

It should be noted carefully that most of the flux induced in the head pole pieces does not thread the head coil. Only that single pair of magnetic poles which straddles the read-head gap produce useful flux that threads the coil. The observation that, at any instant in time, only a single half-wavelength, or digital bit cell, is producing output signal greatly facilitates the understanding of the maximum attainable signal-to-noise ratio discussed in the next chapter. Notice also that only a fraction of the tape flux threads the coil. Some is lost in the space between the tape and the head top surface; this corresponds to the spacing loss. Not all the flux entering the head threads the coil because some leaks across the gap; this flux corresponds to the fact that the read head is not 100% efficient.

When media are vertically magnetized, the phase relationships are changed by 90°. For reproduce heads above and below the medium, the output-signals phases are given by $(-\sin kx)$ and $(+\sin kx)$, respectively; it follows that, in this case, the reproduce-signal processing need not include a 90° phase compensator. In reality, it is usually found that the output signal corresponds to neither longitudinal nor vertical magnetization exactly but rather to a magnetization inclined at some angle θ to the horizontal; accordingly, a phase correction of $90° \pm \theta$ is required in order to preserve precisely the timing or phase information.

6.4 Reproduce-Head Efficiency

Figure 6.6 shows the electrical equivalent circuit of the reproduce head shown in Figure 6.5. Note that since the tape produces a constant flux, the electrical circuit contains a constant current generator. The flux in the read head divides into parallel paths, crossing the gap and threading the coil, respectively; therefore, the electrical circuit has two parallel resistors.

The efficiency of a reproduce head is defined to be that

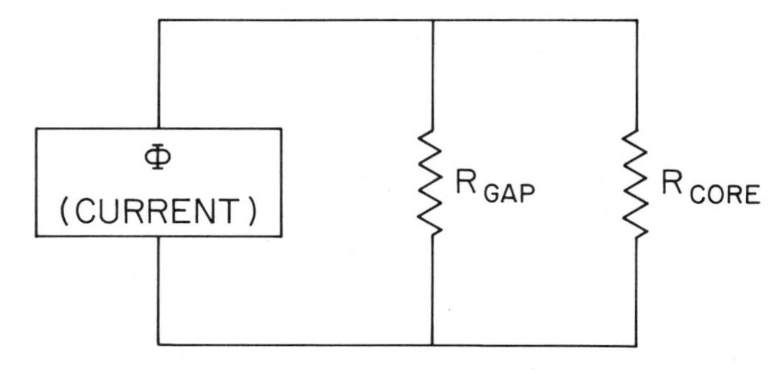

Fig. 6.6. The magnetic circuit equivalent of a read head.

fraction of the flux entering the head that threads the coil. In the electrical equivalent circuit, the analogous quantity is the current flowing through the resistor, which corresponds to the core reluctance. By applying Kirchhoff's laws, it is found that

$$\text{Eff} = \frac{R_g}{R_c + R_g} \tag{6.10}$$

which is precisely the same expression that was given in Chapter 4 for the efficiency of the writing, or record, head.

This correspondence is not fortuitous. It may be demonstrated that, no matter what the materials, design, or construction of a head, the writing efficiency equals the reading efficiency provided only that its behavior is linear. As long as no part of the head saturates, it does not matter what linear hysteresis or eddy current losses or fringing flux paths the head may have, the two efficiencies are identical. The most general explanation of this fact invokes the Reciprocity Principle, which is discussed in the next section.

6.5 The Reciprocity Principle

The Principle of Reciprocity applies to all linear systems, but it is of particular utility and convenience in the analysis of the reproduce process in magnetic recording systems. Before dealing with the reproduce process, however, the generality of reciprocity will be illustrated by two examples taken from mechanics and electromagnetic radiation, respectively.

Suppose that a force vector is applied to point 1 of an elastic structure with the result that a displacement vector is observed at point 2. Reciprocity guarantees that, if the same magnitude force is applied at point 2 in the direction of the original displacement vector, the identical magnitude displacement will be observed at point 1 in a direction parallel to the original force vector. Note that reciprocity links the forces to the displacements and not forces to forces nor displacements to displacements.

As a second example, suppose that a local television transmitting station has 100 A, at 100 MHz, flowing in its antenna and that, some miles away, an open-circuit voltage of 10 microvolts is measured on the terminals of a domestic television antenna. Now suppose that 100 A is imposed, at the same frequency, on the domestic television antenna. Reciprocity asserts that exactly 10 microvolts of open-circuit voltage would then appear on the terminals of the transmitting antenna. Again note that the connection is between voltage and current and not voltage–voltage or current–current.

When a writing head is driven by a current in its coil, it produces flux above the gap. Conversely, when the same head is used in the reading process, the permanently magnetized tape produces flux in the coil. In Chapter 4, the fact that a permanent magnet may be represented by Amperian, or hypothetical, currents, and that these currents yield the **B** field everywhere, was discussed. Accordingly, a permanently magnetized tape may be represented, or replaced, by current bearing coils.

Figure 6.7 shows a head with its normal coil and another coil, above the gap, that represents a small element of the tape. In the initial experiment, current flows in the normal coil producing flux-linkages (flux times turns) in the tape coil. In the reciprocal experiment, the same current flows in the tape coil and reciprocity asserts that the identical flux linkages thread the head coil. The connection is between current and flux linkages and not current-current nor flux linkages-flux linkages.

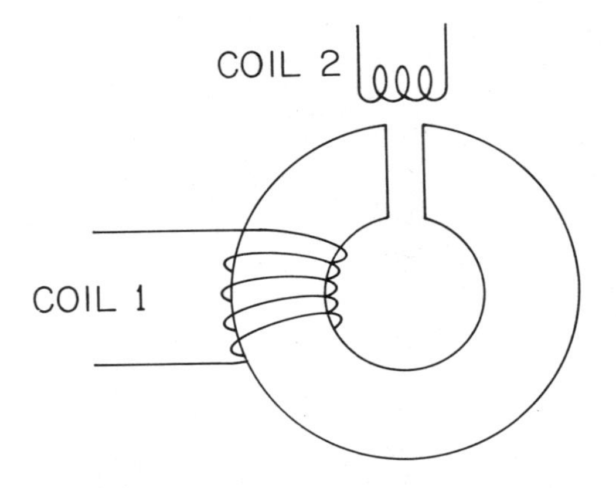

Fig. 6.7. Mutual inductance and reciprocity.

Electrical engineers will recognize that this discussion is tantamount to stating that the coefficients of mutual inductance of the two coils are identical; they have the same flux linkages no matter which carries the current. This idea, carried through the several logical steps listed below, leads to a result of great utility. Suppose current I flows in coil 1. The flux linkages, $d\Delta$, in coil 2 are

$$d\Delta = N_2 \mathbf{H}_2 \cdot d\mathbf{A}_2 \tag{6.11}$$

Replace coil 2 with magnetic material having the same magnetic moment, $\boldsymbol{\mu}$,

$$\boldsymbol{\mu} = N_2 I \, d\mathbf{A}_2 = \mathbf{M}_2 \, dV_2 \tag{6.12}$$

or, rearranging,

$$d\mathbf{A}_2 = \frac{\mathbf{M}_2}{N_2 I} dV_2 \tag{6.13}$$

Now, substitute Equation 6.13 into Equation 6.11,

$$d\Delta = \frac{\mathbf{H}_2 \cdot \mathbf{M}_2}{I} dV_2 \tag{6.14}$$

so that, integrating over all space and dropping subscripts,

$$\Delta = \int \mathbf{h} \cdot \mathbf{M} \, dV \tag{6.15}$$

where $\mathbf{h} = \mathbf{H}/I$. This expression is called the Reciprocity Integral.

In the Reciprocity Integral, Δ is the reproducing-head flux linkages, **h** is the vector magnetic field that would be experienced at a point in the tape if a unit test current (1 Abamp = 10 A) were to flow in the reproducing-head coil, and **M** is the tape remanent magnetization vector. Vectors **h** and **M** are multiplied as a scalar, or inner product,

$$\mathbf{h} \cdot \mathbf{M} = h_x M_x + h_y M_y + h_z M_z \qquad (6.16)$$

The great advantage of the Reciprocity Integral method of computing the reproducing-head flux linkages is that a three-dimensional integral over the volume of the tape only need be evaluated because the unit head fringing field, **h**, is already known. A more conventional approach, which involves finding the exact flux flow at all parts in the reproducing head, is rarely undertaken, even today, because of the complexity in matching the boundary conditions of the gapped head. Moreover, the Reciprocity Integral may be applied immediately to any magnetization waveform such as sine, triangular, and square waves.

It is of great importance to note that the Reciprocity Integral, like all integrals, cannot be reversed. An infinite set of magnetization patterns with different waveshapes and different directions of magnetization can all yield the same reproducing-head flux linkages. It is not possible, therefore, to deduce a unique magnetization pattern in the medium by measurements of the read-head output waveform. Because the external field of a tape is obtained by integration over the magnetic poles, it is, likewise, not possible to deduce unique magnetization patterns from any set of measurements of the medium's external field.

6.6 Read-Head Gap Losses

When a tape of width W is longitudinally magnetized according to the generalized sinusoid

$$M_x = M_0 e^{-jkx} = M_0(\cos kx - j \sin kx) \qquad (6.17)$$

the Reciprocity Integral becomes,

$$\Delta = M_0 W \int_d^{d+\delta} \left[\int_{-\infty}^{\infty} h_x e^{-jkx}\, dx \right] dy \qquad (6.18)$$

The bracketed part is identically the Fourier transform of the horizontal component of the unit head field, **h**.

The horizontal component, h_x, for three cases of interest are discussed below. First, consider the case of the (almost) zero gap, where, ignoring constants of proportionality,

$$h_x = \frac{y}{x^2 + y^2} \qquad (6.19)$$

with Fourier transform,

$$F(h_x) = e^{-ky} \qquad (6.20)$$

Upon performing the integration of Equation 6.18, the result has the same form as Equation 6.5,

$$\Delta = \left[4\pi M_0 W N \left(\frac{1 - e^{-k\delta}}{k} \right) e^{-kd} \sin kx \right] \qquad (6.21)$$

Secondly, consider the Karlquist head field,

$$h_x = \tan^{-1} \left(\frac{yg}{x^2 + y^2 - g^2/4} \right) \qquad (6.22)$$

with Fourier transform,

$$F(h_x) = e^{-ky} \left(\frac{\sin kg/2}{kg/2} \right) \qquad (6.23)$$

and reproducing head flux linkages,

$$\Delta = \left[4\pi M_0 W N \left(\frac{1 - e^{-k\delta}}{k} \right) e^{-kd} \sin kx \right] \left[\frac{\sin kg/2}{kg/2} \right] \qquad (6.24)$$

Finally, although no closed form is known for the exact fringing field of a head, a good approximation for the reproducing-head flux linkages of an actual head is

$$\Delta = \left[4\pi M_0 WN \left(\frac{1 - e^{-k\delta}}{k} \right) e^{-kd} \sin kx \right] \left[\frac{\sin kg/2}{kg/2} \right] \left[\frac{1.25y^2 - g^2}{y^2 - g^2} \right] \quad (6.25)$$

The extra multiplicative terms which appear in these spectra are called the gap-loss expressions. They represent, physically, the fact that the spatial resolution of the read head is determined by its gap length. The gap-loss spectra are shown in Figure 6.8.

The Karlquist approximation has its first gap null, where no reproducing-head flux or voltage can be reproduced, at a wavelength exactly equal to one gap length. This null may be imagined to occur because the pair of poles that straddle the gap are both of the same magnetic polarity. For a real head, the first gap null occurs at a wavelength equal to 112% of the gap length. Thus given a head with a perfect, optically measured, 50 microinch gap, the first gap null should be found experimentally at a wavelength of 56 microinches. In practice, the first null will usually be found at even longer wavelengths because of various faults in the gap geometry, such as gap-edge rounding, which make the gap effectively longer.

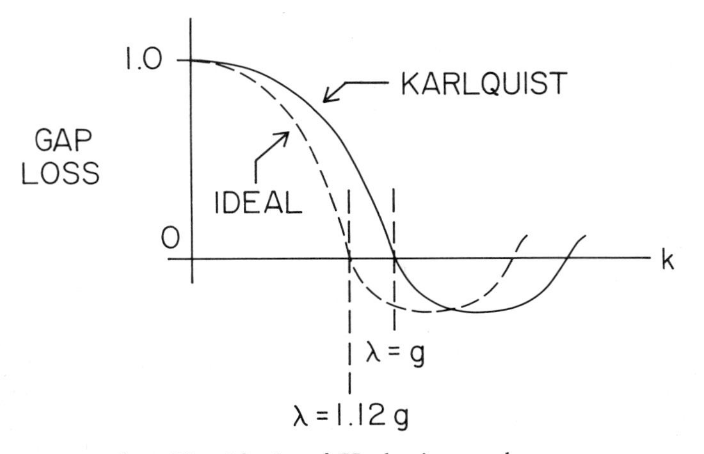

Fig. 6.8. The ideal and Karlquist gap-loss spectra.

Because the two-dimensional fields under consideration have the properties of Hilbert transforms, exactly the same gap-null expressions hold for any direction of tape magnetization.

When the read-head gap length is equal to one-half the wavelength, the gap loss is approximately $2/\pi$, which equals 0.64 or -4 dB. In Chapters 8, 9, and 10, listings are given of the principal parameters of audio and instrumentation recorders, video recorders, and digital recorders, and it may be seen that, in every case, the read-head gap length is about equal to one-half the shortest, or upper bandedge, wavelength used on the medium. The smaller the gap, the lower the head's efficiency and the lower the gap loss. It appears that system designers usually find the optimum read-head gap length to be that which gives a 4 dB gap loss.

6.7 Long-Wavelength Response

Consider the reproducing-head flux linkage when the tape is recorded with a very long wavelength. Now the Fourier transform in Equation 6.18 becomes simply a line integral, thus

$$\int_{-\infty}^{\infty} h_x e^{-jkx}\, dx \rightarrow \int_{-\infty}^{\infty} h_x\, dx \qquad (6.26)$$

According to the analysis of head fields given in Chapter 4, this integral is exactly equal to the mmf across the gap. In reality, however, the line integral must be precisely zero, because the trajectory of the tape does not thread the reproducing-head coil.

Figure 6.9 shows the complete head and not just the region adjacent to the gap. At each end of the head, the horizontal field component is opposite to that over the gap. The area beneath the complete curve, that is, the line integral, is zero. The conclusion is that very long wavelengths cannot be reproduced. In Figure 6.10, the read-head flux spectrum is depicted showing the zero dc response, and at wavelengths

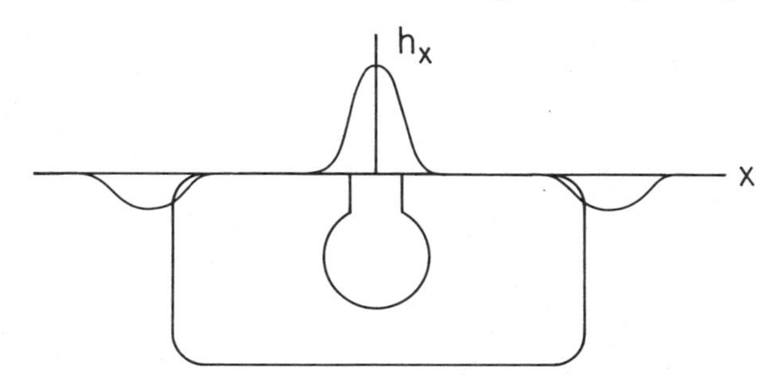

Fig. 6.9. The longitudinal field component above a complete head.

approximately equal to the head's outer dimensions, a series of so-called head bumps. For a head of length L and depth D, the head bumps are found at wavelengths close to $\sqrt{LD}$.

A reproducing head should, therefore, be considered as a spatial, or bandwidth, filter. The lower bandedge is governed by the head's overall dimensions and the shortest wavelength by its gap length. In thin-film heads, the head bumps are usually found in the middle of the spectrum and, by proportioning the thin-film head correctly, modest increases (+4 dB) in the output spectrum can be achieved.

It should be borne in mind that, while the Reciprocity Integral shows the x axis integration extending from $\pm\infty$, this is merely a mathematical artifact. In physical reality, the reproducing-head flux comes only from the half wavelength or bit cell that straddles the gap.

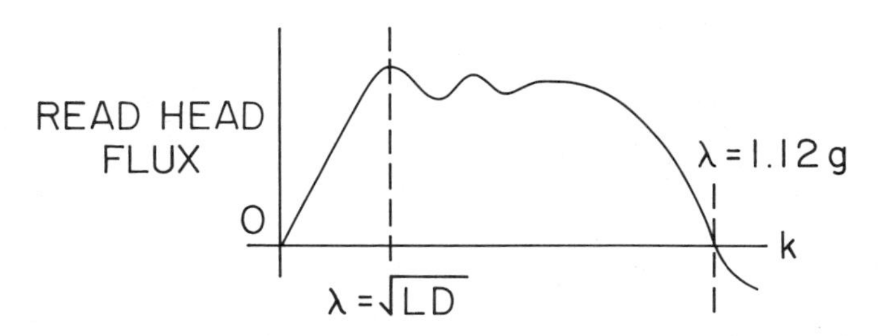

Fig. 6.10. A reproduce spectrum showing the zero dc response and long-wavelength bumps.

6.8 Azimuth and Other Losses

The phase fronts of magnetization across the track width are determined by the trailing edge of the writing-head gap. When, however, the reproduce-head gap is not exactly parallel to the phase fronts, the output flux and voltage are reduced by yet another wave-number-dependent term, called the azimuth loss.

A plan view of the track is shown in Figure 6.11; the difference in the write and gap orientations, called the azimuth error angle, is θ. It is clear that the phase of the reproduce signal changes continuously across the track by an amount $kW\theta$ radians. This results in a reduced output signal, given by

$$E = E(0)\left(\frac{\sin kW\theta/2}{kW\theta/2}\right) \tag{6.27}$$

where $E(0)$ is the output with no azimuth error.

The similarity of this expression to that of the Karlquist head gap-loss expression is self-evident. Note that the first azimuth null occurs when $\lambda = W\theta$, corresponding to 2π radians of phase difference across the track.

Azimuth losses are used to great advantage in consumer

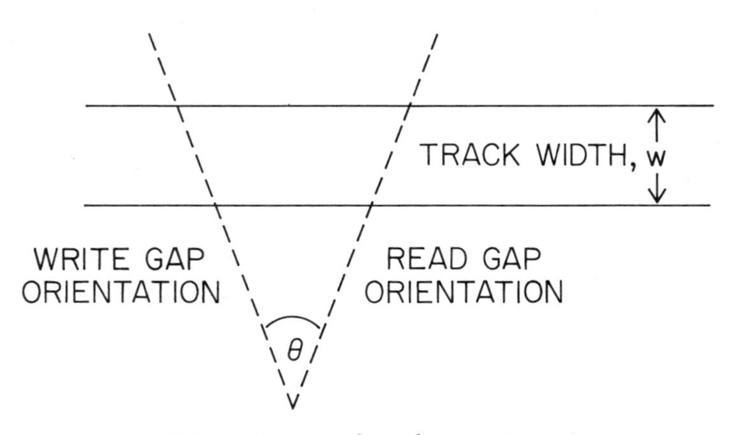

Fig. 6.11. Write- and read-gap orientations.

video recorders to reduce side-reading effects in the high track density recording used. This technique, called the slant-gap recording, is discussed in Chapter 9. In most of the instrumentation recorders described in Chapter 8, provision is made to adjust the reproduce-head azimuth angle to match that of the write head. With typical track widths and shortest wavelengths of 40 mils (40×10^{-3} inches) and 60 microinches, respectively, an azimuth error angle of only 6 minutes of arc is sufficient to reach the first azimuth null; an error angle of 3 minutes of arc yields a -4 dB azimuth loss. Clearly, great mechanical precision is needed in wide-track, short-wavelength recording.

A closely related loss occurs when the gap edges are not perfectly straight and two different heads are used for writing and reading. Gap edges may not be straight for a wide variety of reasons, including bending during lapping and polishing operations and thermal distortion during glass bonding or annealing. Suppose that the deviations from straightness of a head gap can be represented as a Gaussian distribution. Further, suppose that the write and read heads have different and uncorrelated distributions characterized by standard deviations σ_w and σ_r. The reduced output voltage is

$$E = E(0) \exp\left(\frac{-k^2(\sigma_w^2 + \sigma_r^2)}{2}\right) \tag{6.28}$$

where $E(0)$ is the output with straight gaps.

The importance of this effect in short-wavelength recording may be judged by noting that, if the write and read heads have equal root mean square deviations of 0.1λ only, the loss in signal is more than 3 dB. For most short-wavelength recorders, λ is about 40 microinches, and the permitted deviation from absolute straightness is only 4 microinches, or 1000 Å. Compounding this problem is the fact that it is extremely difficult to measure gap straightness to such accuracy; 4 microinches is well below the resolution limits of optical microscopy.

6.9 The Demagnetization–Remagnetization Cycle

When the medium is first magnetized, it is in the close proximity of the write head, whose highly permeable polepieces image out the longitudinal component of the demagnetizing field, particularly in the all important surface, or shallow, layers. Thereafter, as the medium leaves the write head, the demagnetizing field in the tape or disc builds up to its maximum level. At short wavelengths, as can be seen from Equation 6.2, the peak demagnetization factor on the tape surface is 2π and the corresponding magnetization may be determined by finding the intersection of a shear line, of slope $\tan^{-1}(1/2\pi)$ and the M–H loop of the medium. Note that this common application of the shear line is but a graphical method for solving two equations simultaneously; $H_d = -NM$ and the M–H loop.

Now, when the tape approaches the highly permeable reproducing head, the longitudinal component of the demagnetizing field is, again, imaged out. The precise effect of the reduction to zero of the demagnetizing field can be found graphically from the remanence curve, M_r–H, as is shown in

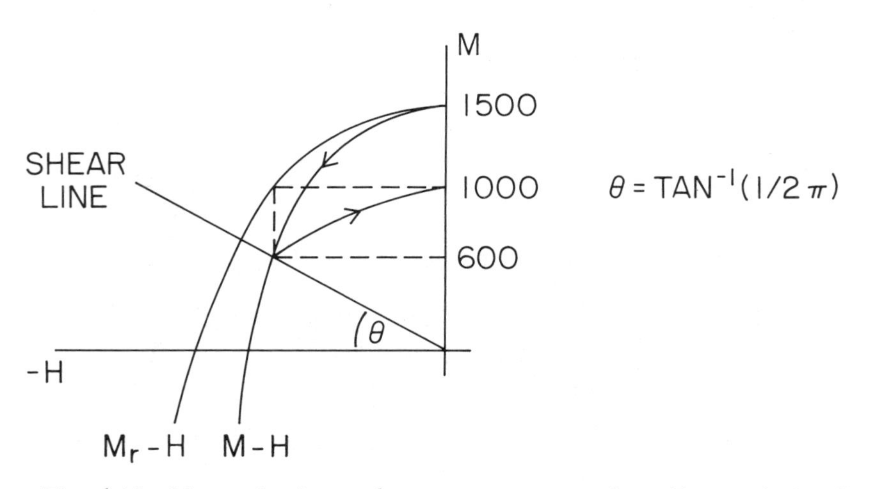

Fig. 6.12. Magnetization and remanence curve shear-line analysis of the demagnetization–remagnetization cycle loss.

Figure 6.12. By definition, the remanence curve shows the magnetization remaining when the total field is reduced to zero. As is shown in Figure 6.12, the magnetization increases, or recoils, as the demagnetizing field is imaged out. For a standard γ-Fe_2O_3 tape magnetized by the write head to a $4\pi M$ value of 1500 G, the magnetization in free space drops to about 600 G and recoils, when close to the read head, to about 1000 G. This latter value is clearly the proper value to use in reproducing-process calculations, because the output flux and signal are determined by it. For other recording media, the demagnetization–remagnetization cycle can be determined quickly by the same method using their M–H and M_r–H loops.

It is obvious that the higher the coercive forces, the lower the reductions in magnetization due to the demagnetization–remagnetization cycle. Figure 6.13 shows the cycle loss, in decibels, versus the intrinsic coercive force of cobalt-surface γ-Fe_2O_3 modified tape media. Very little further reduction in the cycle loss occurs once ${}_mH_c$ is greater than 1000 Oe. Practically, there is rarely any advantage in having ${}_mH_c$ greater than about two-thirds of $4\pi M_r$.

Finally, it may be noted that the reading process in a recorder is, in every way, analogous to the processes occurring in any other electric generator or dynamo where permanent magnets move past a pole structure bearing output coils. It follows that the same considerations of recoil energy, which

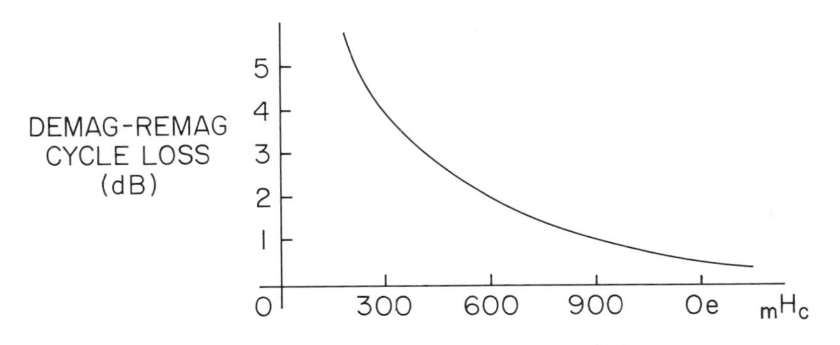

Fig. 6.13. Demagnetization–remagnetization cycle losses versus coercivity.

are considered in discussing the maximum power output of generators, apply equally well to magnetic recording.

Exercises

1. Show how $\exp[-kd]$ becomes $\dfrac{-55d}{\lambda}$ dB.
2. Are the vertical fields above and below a tape in free space, in phase, 90° out of phase or 180° out of phase with a sinusoidal perpendicular magnetization pattern?
3. The output voltage of a ring head is 90° out of phase with a recorded longitudinal magnetization pattern. What corrective action does this mandate?
4. Give an expression for the read-head efficiency.
5. Only a small part of a tape contributes, at any instant, to the output voltage. Which part of the tape is this?
6. How do the impedance, signal power, and noise power of a read head vary with N, the number of turns on the coil?
7. Give the phases, relative to a recorded sine wave of longitudinal magnetization, of (a) the longitudinal field below the tape, (b) the read-head flux, and (c) the output voltage on the coil.
8. What is the magnetostatic, or self-energy, of a uniformly magnetized sphere of volume, V, and magnetization, M?
9. What dimensions of a read head govern the spatial band-pass of that head?
10. Why do all the read heads have their first gap nulls at wavelengths longer than $1.12g$?
11. It is not possible to magnetize uniformly any body that is not an ellipsoid of revolution. Why is this true?
12. Does the efficiency of a read head relate the coil flux to the flux (a) entering the head or (b) leaving the medium?

Further Reading

Duinker, S., and Guerst, J. A. (1964). Long wavelength response of magnetic reproducing heads with rounded outer edges. *Philips Res. Repts.* **19**, 1–28.

Guerst, J. A. (1963). The reciprocity principle in the theory of magnetic recording. *Proc. IEEE* **51**, 1573–1577.

Mallinson, John C. (1966). Demagnetization theory for longitudinal recording. *IEEE Trans. Mag.* **2**, 233–235.

Wallace, R. L., Jr. (1951). The reproduction of magnetically recorded signals. *Bell Syst. Tech. J.*, in *Introduction to Magnetic Recording* (White, R. M., ed.). IEEE Press, New York.

Chapter 7

Noise Processes and Signal-to-Noise Ratios

7.1 Introduction

Noise occurs in all physical systems because of statistical variations from point to point in space, or from instant to instant in time, of some quantity. If everything were precisely uniform, a condition abhorred by nature and denied by quantum theory, there would be no noise. In Chapter 4, the Johnson noise of a resistor was discussed; this noise arises because, statistically, unequal numbers of electrons move toward the resistor terminals at any instant. The noise generated by the imaginary, or lossy, part of a reproducing head's impedance is due to statistical uncertainties, caused by thermal activation, in the motion of the domain walls.

In magnetic recording systems, there are three types of noise to be considered: tape noise, reproducing-head noise, and electronics noise. Tape noise, which is predominant in well-designed recorders, will be analyzed exhaustively in this chapter. Reproducing-head noise, which is expected to become increasingly significant in future recording systems with narrower track widths, was treated in Chapter 4 and needs no further discussion. Electronics noise originates in the first stages of electronic amplification of the reproduced signal; it is almost negligible in most systems and merits no further mention here.

7.2 Additive and Multiplicative Noises

There are two different noises which come from the media. The first is purely additive and comes from the processes in the tape that follow Poisson's or random statistics. When Poisson's statistics are not obeyed, multiplicative, or, as it is often called, modulation, noise arises. The two kinds of noise are shown in Figure 7.1. Notice that, whereas additive noise is present at the same level all of the time, modulation noise varies in level with the signal.

Random statistics are perhaps most familiar in radioactive decay, where it appears that the probability of decay of any given atom is exactly the same as that of any other atom. Another example occurs in gambling; the probability of a given number appearing in a fair game is the same as that of any one of the other numbers.

Additive noise is most important in very high signal-to-noise ratio recorders, such as analog audio and instrumentation recorders. It is due to the particulate, or granular, nature of the recording medium. Given Poisson's statistics, so that there is an equal probability of finding a particle at any point within the coating, and that its direction of magnetization is random, the theory is well-established and is outlined in this chapter.

Multiplicative noise is most important in low signal-to-noise recorders, such as analog video and digital recorders. It is due to deviations from random statistics, such as, for example, local clumping of particles due to poor dispersion and the formation of nonrandom patterns of particle magnetiza-

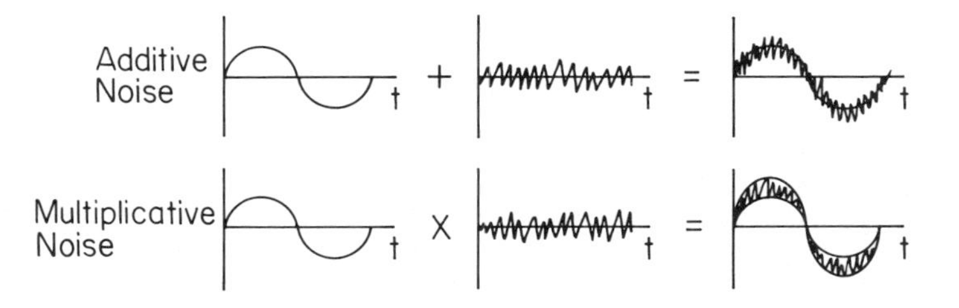

Fig. 7.1. Additive and multiplicative noises.

tions due to their magnetic interaction fields. When the statistics of the nonrandom process are known, there is little difficulty in working out the modulation noise. Generally, however, the nonrandom statistics are not known and very little useful analysis can be performed. Note that the deficiency lies in not knowing the physics and not in the subsequent analysis. Multiplicative noise is not treated further here.

7.3 Addition of Noise Powers

Consider two signals, E_1 and E_2, which are added together. The mean value, or time average, of their sum squared is

$$\overline{(E_1 + E_2)^2} = \overline{E_1^2} + \overline{E_2^2} + \overline{2E_1E_2} \tag{7.1}$$

where the bar above means time average. When the two signals are uncorrelated, that is to say when they bear no statistical relationship to each other, the cross-product term, $2E_1E_2$, has a mean value of zero. In this case, the square of the sum is equal to the sum of the squares. If the signals, E_1 and E_2, are imagined to be voltages applied to a hypothetical one-ohm resistor, then E_1^2 and E_2^2 may be regarded as powers and the power of the sum is the sum of the powers. Because noise comes from random, uncorrelated processes, the rule, known to all electrical engineers, is that the noise powers are added rather than noise voltages. To avoid confusion, hereafter in this book, all signals and noises will be treated as mean powers. For a sine wave of voltage with peak amplitude E, the mean, or time averaged, power is $E^2/2$.

The expression given previously for the Johnson noise of a resistor may be rewritten,

$$E_n^2 = \text{NPS}(f) = 4kT\,R\,\Delta f \tag{7.2}$$

where NPS(f) is the noise power spectrum (watts/Hz), which is a function of frequency. The NPS of a resistor is depicted in Figure 7.2, where the NPS is constant from very low frequencies to very high frequencies. The total noise power, P_n, is obtained by integration of the NPS, thus

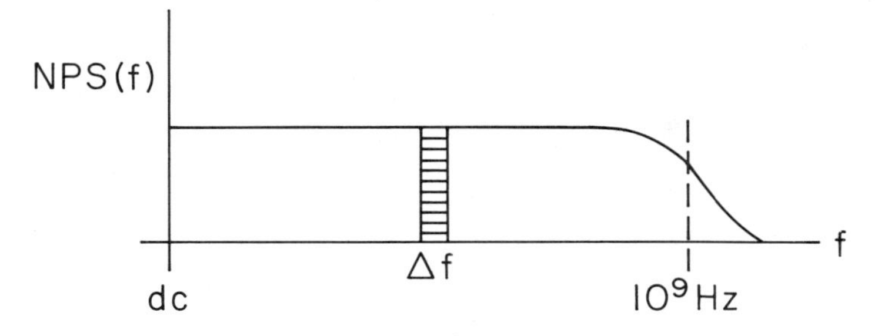

Fig. 7.2. A white-noise power spectrum.

$$P_n = \int \text{NPS}(f)\, df \tag{7.3}$$

Note that because the noise at a particular frequency is uncorrelated with that at any other frequency, the integration follows the rule: add the noise powers.

When dealing with noise power spectra, special care should be taken to understand clearly what frequency increment is being used. Noise power spectra are given both in increments of a fixed number of cycles, for example, 100 Hz, or in increments that are proportional to frequency, such as octaves and decades.

7.4 Tape Noise Total Power

As each magnetic particle in the medium is transported past the reproducing head, it produces a tiny di-pulse of power, as is shown in Figure 7.3. When the particles are magnetized and positioned at random, the power pulses of every particle add incoherently and the total noise power can be calculated easily using reciprocity.

Suppose each particle, of volume v, has magnetic moment $\mu = M_s v$. The mean power, dP, from each particle is, by Faraday's law,

$$dP = \frac{1}{2}\left(\frac{d\phi}{dt}\right)^2 = \frac{1}{2}\left(v\,\frac{d\phi}{dx}\right)^2 \tag{7.4}$$

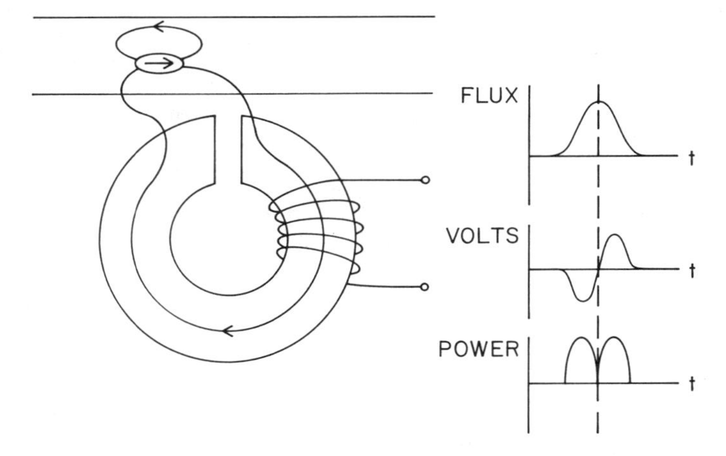

Fig. 7.3. The noise flux, voltage, and power pulses produced by a single magnetic particle.

Now, by reciprocity, the flux from each particle is

$$\phi = \int \mathbf{h} \cdot \mathbf{m} \, d \, \text{vol} = \mathbf{h} \cdot \boldsymbol{\mu} \tag{7.5}$$

so that, when the particles are longitudinally oriented,

$$dP = \frac{v^2 \, \mu^2}{2} \left(\frac{dh_x}{dx} \right)^2 \tag{7.6}$$

If n is the number of particles per unit volume, the total noise power, P, is

$$P = V^2 \frac{\mu^2 n}{2} \int \left(\frac{dh_x}{dx} \right)^2 dx \, dy \, dz \tag{7.7}$$

where the integration is taken over the total volume of the tape. When the field from an (almost) zero-gap length read head is substituted in this equation, the result is

$$P = (10^{-8})^2 4\pi\mu^2 nWV^2 \left(\frac{\delta(d + \delta/2)}{d^2(d + \delta)^2} \right) \tag{7.8}$$

where P is mean watts in a one-turn, 100% efficient read head, μ is the particle's dipole moment in emu, n is the particle density in cm^{-3}, W is the track width in centimeters, V is the head–tape relative velocity in cm/sec, δ is the coating thickness in centimeters, and d is the head–tape spacing in centimeters.

Notice carefully several points about Equation 7.8 that may, at first, seem to be counterintuitive: more particles per unit volume, greater particle magnetic moments, wider track widths, and smaller head–medium spacings all cause greater noise power. The statistics are such that the more sources, the stronger the sources, and the closer are the sources, the greater is the noise power.

7.5 Tape Noise Power Spectrum

A knowledge of the total tape noise power does not provide the tape noise power spectrum because the integration shown in Equation 7.3 cannot be reversed. Clearly, any number of noise power spectra could be imagined that would all have the same integrated noise power; integrations are, unfortunately and inevitably, irreversible.

In order to derive the NPS, a completely separate and different calculation is required, which is beyond the scope of the discussion in this chapter. The result is

$$\mathrm{NPS}(k) = (10^{-8})^2 4\pi\mu^2 nWV^2 k(1 - e^{-2k\delta})e^{-2kd}\,\Delta k \quad (7.9)$$

where NPS(k) is mean watts in a one-turn head, of 100% efficiency, into a one-ohm load, in a fixed wave-number increment, Δk.

This noise power spectrum is actually a so-called two-sided spectrum, that is, it holds for positive and negative wavenumbers, and every k should be read as the magnitude, $|k|$. Two-sided spectra are often used in electrical engineering to keep track of the phase angle of signals. When two-sided spectra are used, it is no longer necessary to use the $j = \sqrt{-1}$ notation discussed previously for this purpose, because for real signals, the real and imaginary parts of their spectra must be even and odd functions of frequency, respectively.

Note that the expression for the NPS(k) spectrum is very similar to those derived in Chapter 6 for the output-voltage spectra. Signal power spectra are derived from the square of the sum of the voltages, whereas noise power spectra are obtained by adding the particle noise powers. Accordingly, all the exponential factors in the NPS appear squared; for example, the spacing loss appears as e^{-2kd}. When Equation 7.9 is integrated over an infinite bandwidth, $-\infty < k < +\infty$, the total noise power given in Equation 7.8 is, of course, obtained.

It is of extreme interest to compare the NPS(k) with signal power spectra because their relative magnitudes determine the mean signal-power-to-mean-noise-power ratios (SNRs). From Chapter 6, it can be shown that the mean signal power, two-sided spectrum for a 100% efficient, one-turn read head is

$$\text{SPS}(k) = (10^{-8})^2 4\pi^2\mu^2 n^2 W^2 V^2 (1 - e^{-k\delta})^2 e^{-2kd} \quad (7.10)$$

where, again, $|k|$ is to be understood.

The signal power and noise power spectra are shown in Figure 7.4. Note that at long wavelengths, both spectra rise at 6 dB per octave due to the differentiating action of the reproducing head. At intermediate wavelengths, or wave numbers, where the thickness-loss term takes effect, the sig-

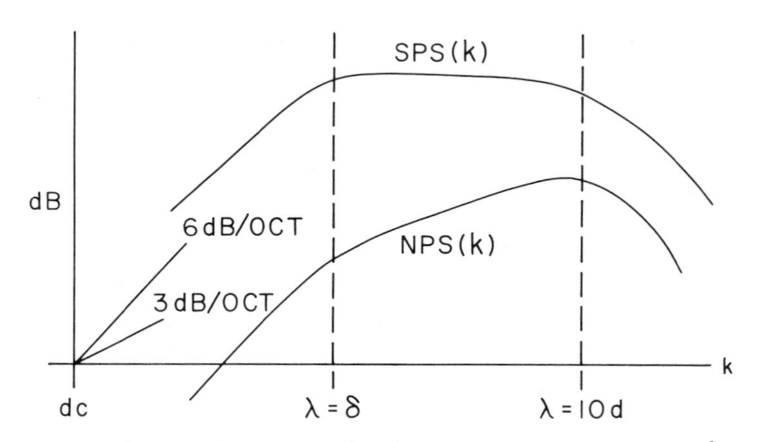

Fig. 7.4. Signal-power and noise-power spectra compared.

nal power spectrum levels off. However, because the noise power spectrum has an extra term, k, the NPS continues to rise at 3 dB per octave. This difference, which occurs at wavelengths comparable to the coating thickness, has had a profound effect on the evolution of magnetic recording systems. This is because it drives the designers constantly to adopt narrower track widths and higher track densities. At shorter wavelengths, say $\lambda = 10d$, the spacing loss reduces both power spectra equally; the spacing losses are equal because the phenomena that determine these power spectra occur within the medium itself.

7.6 Narrow-Band Signal-to-Noise Ratios

By dividing the signal power spectrum by the noise power spectrum, the narrow-band, or slot, signal-to-noise ratio, $(\mathrm{SNR})_\mathrm{n}$, is obtained. Thus,

$$(\mathrm{SNR})_\mathrm{n} = \frac{\mathrm{SPS}(k)}{\mathrm{NPS}(k)} = \frac{\pi n W \tanh(k\,\delta/2)}{k\,\Delta k} \qquad (7.11)$$

In this signal-to-noise ratio, the noise is admitted in a spectral slot of bandwidth Δk centered on the wave number of the signal, k. Since any subsequent filtering, or equalization, operations will change the signal and the slot noise equally, the $(\mathrm{SNR})_\mathrm{n}$ is independent of equalization.

The slot signal-to-noise ratio expression shows some extremely interesting facts. Notice that the magnetic particle dipole moment, μ, does not appear; increases in magnetic moment effect the signal and noise equally. Notice that the head–tape velocity V is not a factor; the particle statistics are not dependent on V. Notice that the head–tape spacing d is absent; provided the additive tape noise remains greater than any other system noise, the attenuation of the signal and noise by spacing losses are equal and are of no consequence.

On the other hand, the narrow-band SNR is directly proportional to the particle packing density, n, and the track width, W. The greater the number of particles in the recorded track, the greater the $(\mathrm{SNR})_\mathrm{n}$.

At long wavelengths, the $(\mathrm{SNR})_n$ becomes

$$(\mathrm{SNR})_n = \frac{\pi n W \delta}{2 \Delta k} \tag{7.12}$$

and it is seen to be independent of wavelength or wave number, k. The thicker the media, however, the greater the $(\mathrm{SNR})_n$, because, again, there are more particles in the recorded track.

At short wavelengths, an approximate form for the $(\mathrm{SNR})_n$ is,

$$(\mathrm{SNR})_n = \frac{\pi n W}{k \Delta k} \tag{7.13}$$

and now the medium's thickness has no influence. This is because the reproducing head cannot, at short wavelengths, sense magnetization deep in the medium. As was noted in Chapter 6, at short wavelengths the majority of the output signal, and, indeed, noise, comes from a shallow surface layer whose depth is governed by the wavelength or reciprocal wave number.

7.7 Wide-Band Signal-to-Noise Ratios

The wide-band signal-to-noise ratio, $(\mathrm{SNR})_w$, is given by dividing the signal power by the total noise power admitted over a wide band width as is indicated in Figure 7.5. Clearly, changes in equalization will change $(\mathrm{SNR})_w$ because the noise is integrated over the range of frequencies or wave numbers distant from that of the signal and the two power spectra are of differing shapes.

In direct analog recording systems, post-equalization is used to make the overall system spectrum flat, that is, to have a constant magnitude over a suitable bandwidth. As is discussed further in Chapter 8, this specific equalization minimizes distortion of the reproduce signal.

Given that the output signal power spectrum is $\mathrm{SPS}(k)$, it is clear that the required post-equalizer power transfer func-

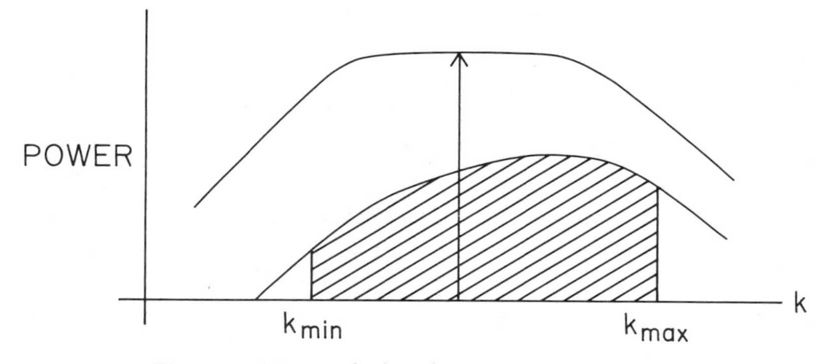

Fig. 7.5. The wide-band signal-to-noise ratio.

tion is proportional to the reciprocal of SPS(k). The overall power transfer function of the write–read processes and post-equalizer cascade, shown in Figure 7.6, is then a constant. Of course, the noise power spectrum is modified also by the post-equalizer. The wide-band signal-to-noise ratio is given by,

$$(\mathrm{SNR})_{\mathrm{w}} = \left[\int_{k_{\min}}^{k_{\max}} \frac{\mathrm{NPS}(k)}{\mathrm{SPS}(k)}\, dk \right]^{-1} \tag{7.14}$$

where $k_{\max}$ and $k_{\min}$ are the upper and lower bandedge wave numbers.

At short wavelengths, $(\mathrm{SNR})_{\mathrm{w}}$ reduces to the approximate form

$$(\mathrm{SNR})_{\mathrm{w}} = \frac{nW\lambda_{\min}^2}{2\pi} \tag{7.15}$$

where n is the density of particles in cm^{-3}, W is the track-width in centimeters, and $\lambda_{\min}$ is the minimum wavelength. Notice again that, at short wavelengths, the coating thick-

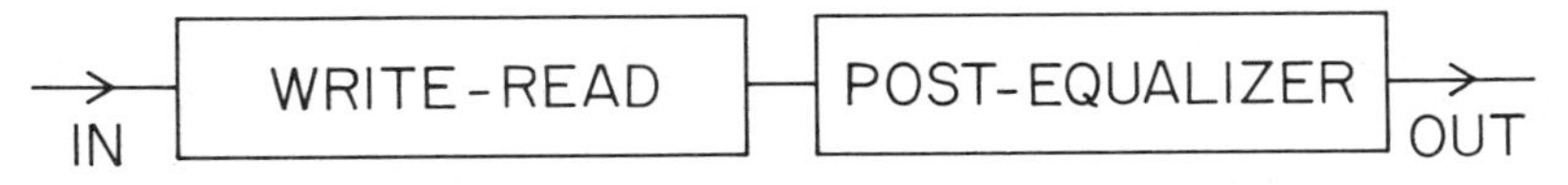

Fig. 7.6. The recorder and post-equalizer cascade.

ness, the particle magnetic moment, and the head–medium velocity do not appear for the reasons discussed in the last section.

The wide-band signal-to-noise ratio expression of Equation 7.15 admits of an extremely simple and illuminating explanation. Figure 7.7 shows the volume of tape being read by the reproducing head at any instant in time. The volume is defined by three dimensions. The width is given by the track width, W. The length is equal to one-half of the wavelength, because only that pair of magnetic poles which straddles the gap produces flux in the reproducing-head coil. The effective reproducing depth is limited by the thickness loss; it is about one-third of the wavelength. The volume of this region is, therefore, given by

$$\text{Volume} = W\frac{\lambda}{2}\frac{\lambda}{3} = \frac{w\lambda^2}{6} \tag{7.16}$$

and, if there are n particles per unit volume, the number of particles being read at any instant is

$$\text{Number} = \frac{nW\lambda^2}{6} \tag{7.17}$$

which is almost identical to Equation 7.15.

It appears, therefore, that the short-wavelength, wide-band, signal-to-noise power ratio of a flat equalized recorder

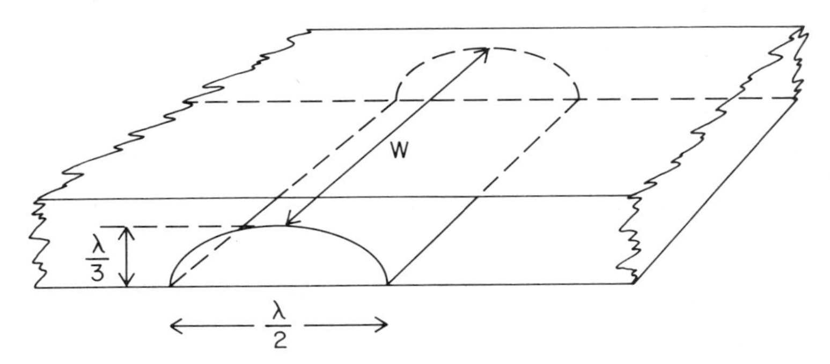

Fig. 7.7. The volume of tape interrogated at any instant in time by the read head.

is equal to the number of particles being read at any instant. That a result of such remarkable simplicity is obtained at the end of a long and complicated analysis serves to stress the statistical origin of the additive tape noise. The signal power is proportional to the number of particles squared. The noise power is the standard deviation squared, or variance, in the number of particles being read and is, by the so-called law of large numbers, proportional to the number. The greater the number, the greater the absolute variance, but the smaller the relative variance. The statistical precision of a measurement increases as greater numbers are involved.

When 10,000 particles are being read at any instant, the signal-to-noise power ratio is 10,000, or 40 dB. For 1000 and 100 particles, the expected signal-to-noise ratios are 30 dB and 20 dB, respectively. This particle counting idea is so simple that it is usually possible to determine the signal-to-noise ratio by mental arithmetic. Like the Karlquist head field approximation, it has the great virtue of simplicity.

The fact that $(SNR)_w$ is directly proportional to the track width but depends on the square of the shortest wavelength permits conclusions to be drawn that dominate the evolution of recorders. There are two possible ways to double the areal density of a recorder: halve the track width or halve the wavelength. Halving the track width reduces the $(SNR)_w$ by a factor of two (-3 dB). Halving the wavelength reduces the $(SNR)_w$ by at least a factor four (-6 dB). It is manifestly obvious that reducing the track width, and concomittently increasing the track density, is the preferred method for increasing the areal density. Accordingly, throughout the history of video recorders, for example, track widths have been reduced by a factor of 20, while linear densities have been reduced by a factor of only four.

The superiority of narrow track widths over short wavelengths is unfortunate for the amateur recording engineer. Halving the wavelength can be accomplished easily by merely halving the head–medium relative velocity; often this requires nothing more than a change in diameter of a belt pulley or capstan. Halving the track width, on the other hand, requires that the write and read heads be changed; this is, of course, beyond the abilities of the tyro. The history of recording is filled with unsuccessful attempts to make recorders with twice the playing time by merely halving the tape speed; it is not the proper procedure.

7.8 Packing Density of Particles

All the signal-to-noise ratio expressions considered in this chapter have been directly proportional to the particle packing density, n. Provided that other system noises are negligible, the particle packing density is the only tape parameter of consequence.

The evolution of improved recording media may be characterized as a gradual increase in n, made possible by using smaller and smaller magnetic particles. In most applications, it is necessary to limit the number, or fraction, of superparamagnetic particles present in the medium. Attempts to size fractionate magnetic particles have, in the main, not been successful. It is, therefore, not possible to remove the tiny superparamagnetic particles. Accordingly, the only way left open to use smaller particles is to use magnetic materials with higher switching energies. This has usually been achieved by increasing the coercivity, but more recently, substantial increases in the magnetization have also been used.

Pure γ-Fe_2O_3 particles, of the type used in audio tapes in the 1950s, were about 40 microinches (1.0 μm) in length and the corresponding density of the particles was approximately 10^{+14} per cubic centimeter. By the mid-1970s, all the video tape manufacturers had adapted cobalt-surface-modified γ-Fe_2O_3 particles, typically of length 20 microinches (0.5 μm); these tapes have a particle density of 10^{+15}. The metallic iron particle tapes now appearing with the rotary head, helical scan recorders discussed in Chapter 9, attain particle densities of nearly 10^{+16} by using particles smaller than 10 microinches (0.25 μm). No doubt, further increases in the packing density of particles will be achieved by increasing the coercivity even more and reducing even further the particle size.

Exercises

1. What change in $(SNR)_w$ follows when only the track width is halved?
2. What change in $(SNR)_w$ follows when only the head–medium speed is halved?
3. The signal power and the noise power of a read head both vary with the number of turns, N, as N^2. What then determines the number of turns actually used in a recording system?
4. Give the dimensions of the volume of the medium interrogated by the read head at any instant in time.
5. If the $(SNR)_w$ is 33 dB, how many particles are being read, on the average, by the read head?
6. Why are noise powers added rather than noise voltages?
7. Suppose that the particle dimensions are halved, but the pigment volume concentration of the medium is not changed. What change in SNR will this cause in a tape noise limited system?
8. Why is the noise power obtained by integrating the noise power spectrum?
9. The phase of the output signal in longitudinal recording lags 90° behind that of the recorded signal. How does this affect the SNR?
10. Show why making a lossy, conventional read head physically smaller reduces its noise.
11. Is the particulate noise power directly proportional to the variance or to the standard deviation of the number of particles being sensed at any instant?

12. Suppose that a novel read head of radically different design is used. Will this change the tape limited narrow band SNR?

Further Reading

Mallinson, John C. (1969). Maximum signal to noise ratio of a tape recorder. *IEEE Trans. Mag.* **5**, 182–186.

Thurlings, L. (1980). Statistical analysis of signal and noise in magnetic recording. *IEEE Trans. Mag.* **16**, 501–513.

Chapter 8

Audio and Instrumentation Recorders

8.1 Introduction

All consumer audio recorders and the majority of instrumentation recorders use ac bias and they are, for this reason, discussed together in this chapter. Although the first magnetic recorders made in the 1930s used dc bias in an attempt to achieve linearity, ac bias was adopted soon after. In an ac-biased recorder, a high frequency, high amplitude ac current is added to the signal current in the write head. Because no complicated modulation scheme is used, ac-biased recorders are often called direct recorders. Figure 8.1 shows the first audio recorder made, in 1948, by the Ampex Corporation.

Instrumentation recorders are very similar to audio recorders in principle; both are linear and equalization is used to limit distortion of the signals. The major differences lie in the number of signal channels and the frequency of operation. Audio recorders generally have but two channels operating at 15–20 KHz; instrumentation recorders often have 28 parallel channels handling 2 MHz signals.

8.2 Linearity of the Anhysteretic Curve

Alternating-current-biased recording is used because it is the cheapest, simplest way known to make a recording system

Fig. 8.1. Alexander M. Poniatoff and his chief engineer, Harold Lindsay, in 1948 with the first Ampex audio recorder, the Ampex Model 100. (Courtesy Ampex Corporation.)

behave as a linear system. As was given earlier in Chapter 5, a linear system is defined mathematically by

$$L[ax(t) + by(t)] = aL[x(t)] + bL[y(t)] \tag{8.1}$$

where L is a linear transformation, a and b are arbitrary scale factors, and $x(t)$ and $y(t)$ are independent signals. In other words, the linear transform of the sum is just the sum of the linear transforms; no higher power or cross products appear.

The anhysteretic curve shown in Figure 8.2 has the approximate form of the error function, which is the integral of a Gaussian, and the central part is almost a straight line. This straight section means that an almost linear relationship exists between the dc field and the anhysteretic remanent magnetization. The linearity of ac-biased recording occurs in a closely related way: the large amplitude ac bias current in the

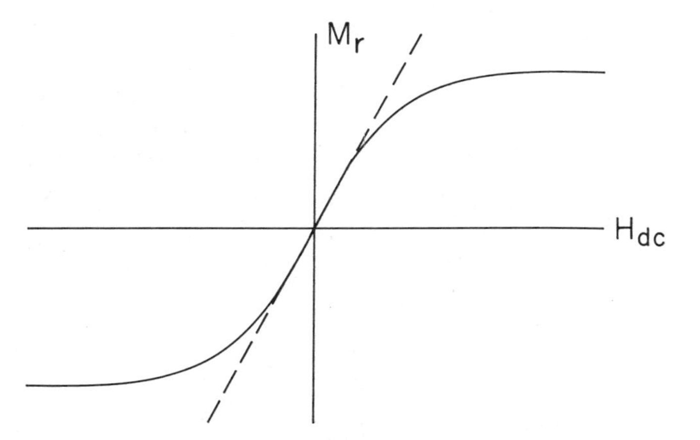

Fig. 8.2. The anhysteretic remanent magnetization curve.

write head provides the energy required to switch the tape particles, and the low-level signal current produces the small field that determines the final remanence. Because both fields decrease in magnitude together with increasing distance from the write-head gap, the process of anhysteresis is actually slightly different. It is called the "modified" anhysteretic process, but the distinction is slight and is not pursued here.

Because the central part of the curve is not exactly straight but curves slightly, the system is not precisely linear. The effect of the slight curvature is to add odd (third, fifth, etc.) harmonic components to a pure sine wave. It can be calculated that 1% and 3% third harmonic components occur when the peak amplitude of the written sine wave reaches 23% and 32% of the maximum anhysteretic remanence, respectively. In order to limit the third harmonic content of the read signal in ac-biased recording, it is therefore necessary to limit the maximum signal level. It follows that the signal is only recorded, or written, using a small fraction of the particles in the medium. All the remaining particles are unsuitably located, with respect to their neighbors and their magnetic interaction fields, to permit low odd harmonic component recording. Unfortunately, all the particles in the medium contribute to the noise, with the result that the signal-to-noise ratio is, inevitably, less than the maximum possible.

Although ac bias is used in the overwhelming majority of magnetic recorders, it is important to realize that the linearity is achieved at the cost of a lowered signal-to-noise ratio. A design trade-off must be made between the lower SNR incurred by ac-biased recording and the electronic signal processing complexity involved in using the modulation schemes, discussed in Chapters 9 and 10, in order to linearize the system.

8.3 AC-Biased Writing Processes

The ac-biased recording process is more complicated to follow than is the unbiased case because both the ac and the signal fields have to be followed simultaneously. When adjusting an ac-biased recorder, both the ac bias current and the signal current levels have to be optimized or set according to some criteria.

Because the ac bias current and write-head fringing fields are much larger (typically a factor of 10) than those of the signal, simplifications exist. The geometry, or spatial configuration, of the writing process is determined almost entirely by the ac field, whereas the signal field merely controls the level of anhysteretic magnetization recorded.

Because the function of the ac bias field is only that of switching the tape particle's magnetization back and forth, it is usually considered that it is the total vector ac field that is of importance, no matter what may be the orientation of the tape particles. When the total field is adjusted so that the contour equal to the coercivity of the medium penetrates fully through the whole coating thickness, as shown in Figure 8.3, the maximum long-wavelength output for all signal levels occurs. Most audio recorders have their ac bias level set in this manner because human hearing has its highest acuity at relatively low frequencies (1000 Hz) or long wavelengths.

At short wavelengths, such a high ac bias level does not yield the best results. For a combination of the reasons dis-

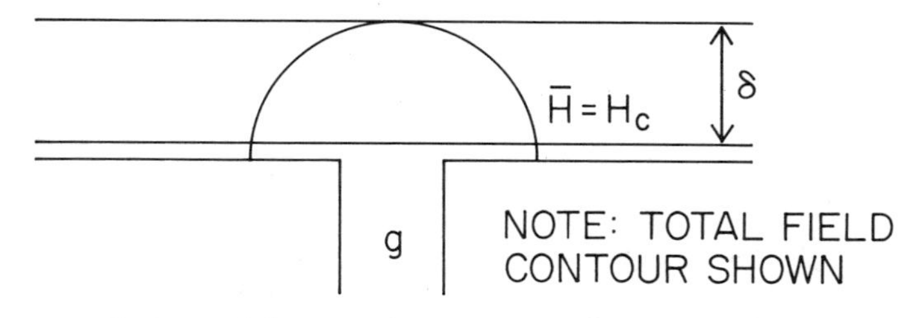

Fig. 8.3. The writing geometry for high ac bias.

cussed in Chapter 5, namely field gradient and phasing errors, the highest short-wavelength output occurs when the ac bias is considerably reduced and the total field contour only partially penetrates the coating as indicated in Figure 8.4. Most instrumentation receivers have their ac bias adjusted to this short-wavelength, high-frequency optimum condition because it yields the best slot signal-to-noise ratios at high frequencies.

The designer of an ac-biased recording system is thus presented with this dilemma: high ac bias gives large and small output signal levels at low and high frequencies, respectively, with low ac bias behaving conversely. The situation is shown in Figure 8.5; note carefully that the variable parameter is the ac bias amplitude and that the abscissa is the frequency, or wave number, of the signal.

In Figure 8.5, four curves are shown. The highest curve is the upper bound, which would only be attained if it were possible to record to the maximum remanence and in phase to the full depth of the coating. The next curve down shows

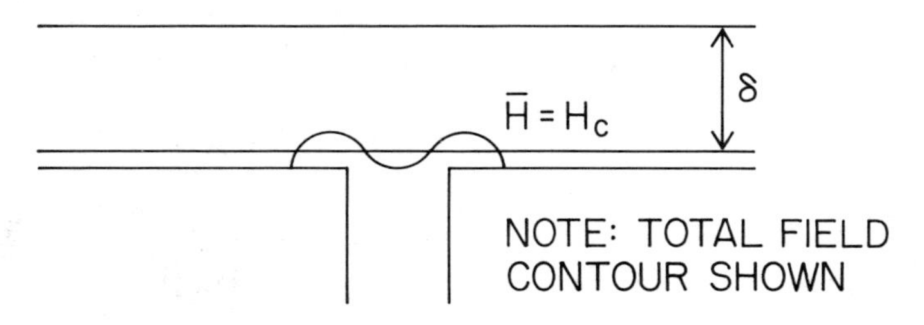

Fig. 8.4. The writing geometry for low ac bias.

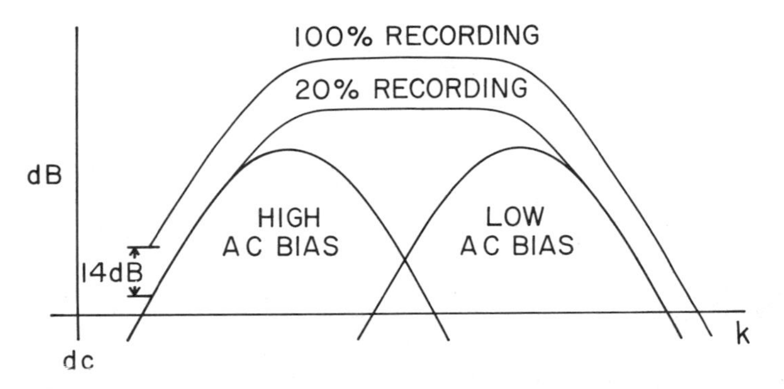

Fig. 8.5. Signal spectra attainable with high and low ac bias.

the spectrum expected if it were possible to record 20% of the maximum remanence and in phase to the full coating thickness. This curve is parallel to the upper curve and at one-fifth (−14 dB) its level. The lower curve on the left-hand side is the output spectrum obtained when the ac bias is large, as in audio recorders; at short wavelengths, the output level is very low due to field gradient and phasing effects. The lower right-hand curve indicates the results obtained with the optimum short-wavelength ac bias method used in instrumentation recorders; although the short-wavelength output is high, at long wavelengths the signal is very low because the depth of recording is so small.

In audio recording, successful attempts to alleviate the ac bias dilemma have been made by using double, or two-layer, coatings. By providing a high coercivity, thin surface layer on a low coercivity coating, it is possible to make tapes for which the long- and short-wavelength ac bias optima are nearly coincident. The extension of this idea to multilayer tapes, which have the coercivities graded through the coating thickness to match exactly the decrease of the head field, appears to be hardly worth the effort involved.

8.4 Amplitude and Phase Equalization

The use of ac bias makes the recorder almost linear. Nevertheless, the output signal still suffers distortion; that is to say, the output signal is not an exact replica of the input signal. This occurs because different frequency, or wave number, components of the signal are reproduced at different amplitudes; moreover, as was mentioned in Chapter 6, the phase of the output signal does not match that of the input signal. Linearity and distortion should not be confused. A linear system may or may not have distortion; a nonlinear system necessarily has distortion.

In order to correct the distortion, specific equalization, or correction, of the amplitude and the phase response of the write–read process is needed. In a linear system, the distortion correction may be accomplished by either pre- or post-equalization; that is, by using filters before or after the write–read process, as shown in Figure 8.6. When more pre-emphasis is used, the system signal-to-noise ratio is higher, but, of course, the odd harmonic distortion is also greater. When the interchange of recordings is required, the division between pre- and post-emphasis must conform to a common standard.

Consider a linear system with input and output signals, $i(t)$ and $o(t)$, respectively. If the Fourier transforms of these signals are $I(\omega)$ and $O(\omega)$, respectively, the frequency-domain transfer function of the system is

$$T(\omega) = \frac{O(\omega)}{I(\omega)} = A(\omega)e^{-j\theta(\omega)} \tag{8.2}$$

where $T(\omega)$ is the (voltage) transfer function, $A(\omega)$ is the amplitude response, and $\theta(\omega)$ is the phase response. In order to remove distortion, the amplitude and phase responses must meet the specific requirements given below.

The amplitude response of the system must be of constant amplitude, or flat, over a suitable range of frequencies, as is shown in Figure 8.7. Output signals must have the same amplitude at all frequencies of importance. In audio recorders, the bandwidth is typically 50 Hz to 20 kHz. In

Fig. 8.6. The pre-equalizer, recorder, and post-equalizer cascade.

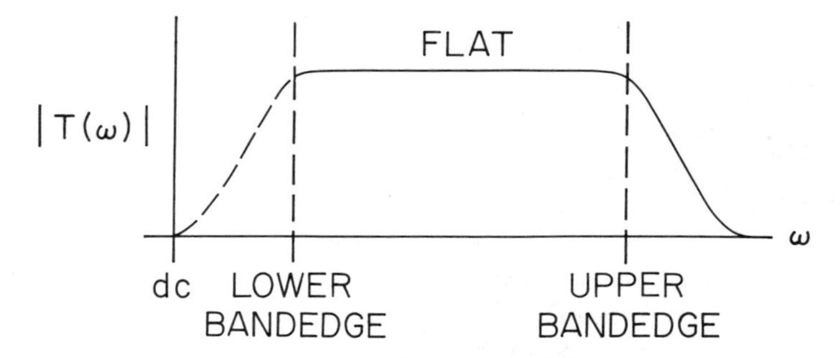

Fig. 8.7. The amplitude spectrum for distortionless recording.

instrumentation machines, 400 Hz to 2 MHz is a frequently used standard. As was observed in Chapter 7, the power transfer function required of a flat amplitude equalizer is, essentially, the reciprocal of the signal power spectrum, SPS(k), of the write–read process.

The phase response required to eliminate phase distortion is shown in Figure 8.8 and it must meet two criteria. First, it must be a straight line over all frequencies of importance. The criterion assures that signals at all frequencies are delayed by exactly the same interval of time; it is accordingly called "constant group delay." Second, the extrapolation of the straight-line part of the phase response must intersect

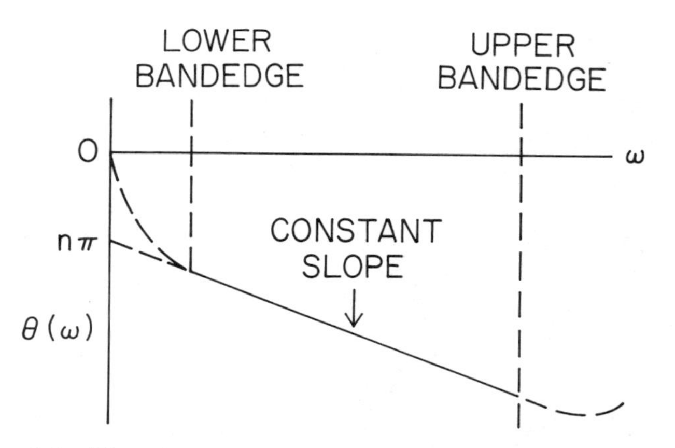

Fig. 8.8. The phase spectrum for distortionless recording.

zero frequency at exactly an integer (0, 1, 2, etc.) value of π radians. The output of a longitudinal recording does not meet this criterion, whereas that of a perpendicular recording does. The phase equalization used in audio recording is usually limited to that of correcting the zero-frequency intercept by integration; the human ear is relatively insensitive to phase distortion caused by deviations from constant group delay. In instrumentation recorders, on the other hand, precise phase equalization is employed because an exact replica output, corrupted only by noise, is a system requirement.

When the relative motion of a tape is reversed, the magnetization sequence on tape is, of course, also reversed. It may be shown that, when the phase response of the reproduce head and post-amplifier cascade, as distinguished from that of the complete system, has a zero-frequency intercept of $n\pi$ radians, the output signal is exactly time reversed. When the zero-frequency intercept is $n\ \pi/2$, the output signal is reversed also but has the opposite polarity. These phenomena are of great convenience in, for example, tape duplication, or dubbing, operations.

8.5 Machine Specifications and Parameters

In Table 8.1, the principal characteristics of two recorders are listed. The audio recorder is a professional analog machine of the type used in recording studios and conforms to National Association of Broadcasters (NAB) specifications. The instrumentation recorder is of the type used in scientific applications, such as military telemetry, and follows Interrange Instrumentation Group (IRIG) specifications.

8.6 Audio and Instrumentation SNRs

In Chapter 7, the wide-band signal-to-noise ratios were elucidated from first principles for the case in which the tape is magnetized to its maximum remanence. In ac-biased recording, however, this yields unacceptably high distortion, and it

Table 8.1
Principal Parameters of Audio and Instrumentation Recorders

Parameter	Audio	Instrumentation
Upper frequency, (kHz)	15	2000
Lower frequency, (Hz)	50	400
Ac bias frequency, (kHz)	100	7500
$(SNR)_w$, (dB)	55	30
Tape speed, (ips)	30	120
Longest wavelength, (inch)	0.5	0.5
Shortest wavelength, (microinch)	150	60
Ac bias criterion	max. long	max. short
Signal criterion	3% third harmonic	1% third harmonic
Write gap-length, (microinch)	250	80
Read gap-length, (microinch)	60	25
Tape type	Co-γ-Fe_2O_3	Co-γ-Fe_2O_3
Tape width, (inch)	0.25	1
Coating thickness, (microinch)	200–500	100–200
Track width, (mils)	80	25
Number of tracks/ channels	2	28

is necessary to record at lower levels. It follows that the expression given in Chapter 7 for the $(SNR)_w$ must be modified to read,

$$(SNR)_w = \frac{nWf^2}{2\pi} \lambda_{min}^2 \tag{8.3}$$

where f is the fraction of the maximum remanence, or of the particles that are suitably positioned for low distortion recording.

Consider the audio recorder with the specifications listed in the previous section. The machine is operated at the 3% third harmonic distortion level, and accordingly, f is equal to

0.32, or about one-third. Substituting $n = 10^{15}$ particles per cubic centimeter, $W = 80$ mils ($80 \times 2.54 \times 10^{-3}$ cm) and $\lambda_{min} = 150$ microinches ($150 \times 2.54 \times 10^{-6}$ cm) into Equation 8.3 yields

$$\text{Audio(SNR)}_w = 5 \times 10^5 \text{ (or 57 dB)} \qquad (8.4)$$

This value is in very close agreement with the experimentally observed value of 55 dB.

For the instrumentation recorder, the third harmonic distortion level is adjusted to 1% only, corresponding to an f value of 0.23. Substituting $n = 10^{15}$, $w = 25$ mils, and $\lambda_{min} =$ 60 microinches gives

$$\text{Instrumentation(SNR)}_w = 8 \times 10^3 \text{ (or 39 dB)} \qquad (8.5)$$

For the instrumentation recorder, the $(SNR)_w$ measured is only 30 dB; most of the discrepancy is due to the fact that the reproducing-head noise is larger than other noises toward the upper bandedge. The relatively high reproducing-head noise is, of course, a consequence of the large size read head needed to handle the long wavelengths and the high frequency of operation.

These excellent agreements between the results of simple theory and experiment are somewhat fortuitous. In reality, the signal power spectrum is considerably lower, particularly at short wavelengths, than those given in Chapter 6; the discrepancy is shown in Figure 8.5. Additionally, the measured noise power spectrum is considerably lower than that derived in Chapter 7; the difference can be attributed to several nonrandom processes. Coincidentally, however, both deviations from simple theory are about the same in magnitude and, therefore, the calculated SNRs turn out to be almost correct.

8.7 Shannon Capacity of Recorders

The Shannon capacity of a linear channel sets a mathematical upper bound to the rate at which information can be transmitted and received error free. It is to be emphasized that it is not a prescription of how to attain the maximum rate in practice but merely a hypothetical limit.

In information theory, information is measured in binary digits, commonly abbreviated bits. The information contained in a message, occurring with probability, p, is defined to be

$$\text{Information} = -\log_2(p) \tag{8.6}$$

This definition follows immediately from the fact that there are only 2^N different binary words of length N bits. The more probable the message, the less is the information it bears. This definition is very similar to that used in statistical mechanics for the thermodynamic quantity entropy, where

$$\text{Entropy} = +\log_e(p) \tag{8.7}$$

As the entropy of the universe inexorably increases, all things become more certain and, thus, contain less information.

Consider a linear channel of bandwidth B Hz. Since there are but two independent properties of a sine wave of given frequency, namely the amplitude and the phase angle, it follows that only $2B$ independent samples per second can be transmitted in this channel. This is called the Nyquist rate, and it is discussed further in Chapter 10. If the mean signal power is S and the mean noise power is N, the number of amplitude levels distinguishable, on the average, is

$$\text{Number of levels} = \sqrt{\frac{S + N}{N}} = \sqrt{1 + \text{SNR}} \tag{8.8}$$

If the amplitude levels are equiprobably distributed, then their probability of occurrence is

$$\text{Probability} = \sqrt{\frac{1}{1 + \text{SNR}}} \tag{8.9}$$

and the information borne by each sample is

$$\text{Information/sample} = \tfrac{1}{2}\log_2(1 + \text{SNR}) \tag{8.10}$$

With $2B$ independent samples per second, it follows that the

Shannon capacity or rate is,

$$C = B \log_2(1 + \mathrm{SNR}) \tag{8.11}$$

Consider the IRIG instrumentation recorder discussed previously, which has a bandwidth, B, of 2 MHz and a $(\mathrm{SNR})_w$ of about 10^3 (30 dB). The Shannon capacity of the machine is

$$C = 2 \times 10^6 \log_2(1 + 10^3) \tag{8.12}$$

which is approximately 20×10^6, or 20 megabits per second. Many such machines are used for recording binary digital data and a common data rate used is 4 megabits per second: this is one-fifth of the Shannon bound. In order to approach the limit more closely, it is believed that more complicated coding, both in the channel code and error detection and correction code senses, is required. These topics, however, will not be examined in this book.

8.8 Ultimate Information Areal Capacity

Consider a tape of unit width, divided, without guard bands between the tracks, into M parallel tracks as shown in Figure 8.9. Clearly, the combined Shannon areal capacity is

$$C(M) = MB \log_2 \left(1 + \frac{(\mathrm{SNR})_w}{M} \right) \tag{8.13}$$

where $(\mathrm{SNR})_w$ is the wide-band signal-to-noise ratio when

Fig. 8.9. The division of tape into parallel tracks.

the full tape width is used for a single track ($M = 1$). For unit head–tape relative speed, the bandwidth B is the reciprocal of the minimum wavelength, λ_{min}. It is obvious from this expression that the areal capacity increases as M increases. The ultimate storage capacity, in bits per unit area, occurs when the number of tracks becomes very large and is

$$\text{Ultimate areal capacity} = \frac{nf^2\lambda_{min}}{2\pi \log e^2} \qquad (8.14)$$

For the instrumentation recorders considered in this chapter ($n = 10^{15}$ particles per cubic centimeter, $f = 0.2$ and $\lambda_{min} = 60$ microinches), the ultimate capacity is, amazingly enough, 10^9 bits per square centimeter. This is an extremely high areal density in comparison with present achievements in digital recording, where figures in the range of 10^6 to 10^7 are the norm.

This discussion makes it plain that, from the point of view of information theory, the signal-to-noise ratios obtained in magnetic recording are much higher than is needed for efficient error-free digital communication. There is good reason to believe that as the development of large-scale integrated (LSI) semiconductor devices continues, magnetic recording system designers will see fit to reduce the track width and, concomitantly, the signal-to-noise ratio and thus achieve considerably higher areal densities in the future.

Exercises

1. What amplitude response is needed for distortionless transmission of signals?
2. What phase response is needed for distortionless transmission of signals?
3. Show why the definition of information in binary digits is information $= -\log_2(\text{probability})$.
4. How can a linear system yield distorted signals?

5. Does the 90° phase shift between input and output signals in longitudinal recording cause (a) noise, (b) distortion, or (c) interference?
6. Why are two-layer audio tapes used?
7. If an audio recorder has a $(SNR)_w$ = 60 dB when 0.1 inch wide tracks are used, how wide must the tracks be made to achieve a $(SNR)_w$ = 90 dB?
8. Why does the use of ac bias to linearize a recorder's response cause a reduction in SNR?
9. What is the power transfer function required of an equalizer in order to make a recorder's amplitude response flat?
10. How does post-equalization change the $(SNR)_w$?
11. When pre- and post-equalization are used to make the recorder have a flat amplitude response, which gives the higher $(SNR)_w$?
12. What is the Shannon capacity of a linear channel of bandwidth 1 MHz and $(SNR)_w$ = 15 dB?

Further Reading

Bertram, H. Neal (1974). Long wavelength ac bias recording theory. *IEEE Trans. Mag.* **10,** 1039–1048.

Eldridge, D. F. (1963). A special application of information theory to recording systems. *IEEE Trans. Audio* **AU-11,** 3–6.

Jorgensen, Finn (1980). *The Complete Handbook of Magnetic Recording.* TAB Books, Blue Ridge Summit, Pennsylvania.

Lathi, B. P. (1968). *Communications Systems.* Wiley, New York.

Lowman, C. E. (1972). *Magnetic Recording.* McGraw-Hill, New York.

Mallinson, John C., and Ferrier, H. (1974). Motion reversal invariance in tape recorders. *IEEE Trans. Mag.* **10,** 1048–1049.

Chapter 9

Video Recorders

9.1 Introduction

The two developments which particularly characterize video recorders are the use of frequency modulation, FM, and rotating heads. Despite early, unsuccessful attempts to use direct recording and fixed heads, the overwhelming majority of current video recorders are FM rotary head machines. Although digital video recorders will appear in small numbers in television studios, it is expected that analog FM recorders will predominate for many years to come. Figure 9.1 shows, in 1956, the world's first successful commercial video recorder, the Ampex VR1000. The machine weighed over 1000 pounds.

9.2 Frequency Modulation

Before discussing why frequency modulation is so well suited to video recording, two closely related analog modulation methods will be discussed.

Amplitude modulation is used in AM radio broadcasts. The signal modulates, that is, multiplies, the amplitude of a relatively high frequency carrier. Thus

$$S_{\mathrm{am}}(t) = f(t)\cos(\omega_c t) \qquad (9.1)$$

where $f(t)$ is the signal and ω_c is the angular frequency of the carrier. In phase and frequency modulation, however, the

Fig. 9.1. Alexander M. Poniatoff in 1956 with the world's first video recorder, the Ampex VR1000. The tall cabinet behind Mr. Poniatoff housed much of the electronics with several hundred vacuum tubes. (Courtesy Ampex Corporation.)

amplitude remains constant and the signal changes the phase angle of the carrier. For phase modulation, PM,

$$S_{\mathrm{pm}}(t) = \text{constant} \cos(\omega_c t + 2\pi\beta f(t)) \tag{9.2}$$

and, for frequency modulation, FM,

$$S_{\mathrm{fm}}(t) = \text{constant} \cos\left(\omega_c t + 2\pi\beta \int f(t)\, dt\right) \tag{9.3}$$

Both cases are called angle modulation; the only difference lies in whether the carrier phase angle change is proportional to the signal or the integrated signal. The constant of proportionality, β, is called the sensitivity (Hz/volt). When the signal frequency is ω_s, the ratio $2\pi\beta/\omega_s$ is called the modulation index.

Figure 9.2 shows an amplitude modulated and a frequency modulated carrier. In the FM case, note that it is very diffi-

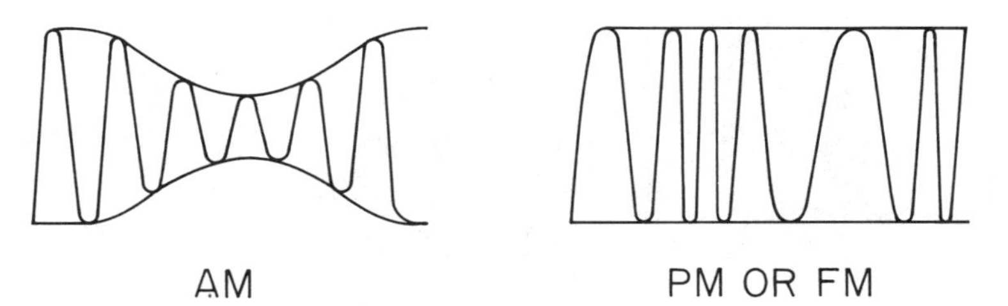

Fig. 9.2. Amplitude modulated, phase modulated, and frequency modulated waveforms compared.

cult to determine the frequency because the waveshape varies from time to time. This difficulty leads to the definition of "instantaneous angular frequency," ω_i, where

$$\omega_i = \frac{d(\text{angle})}{dt} \tag{9.4}$$

In the case of FM,

$$\omega_i = \omega_c + 2\pi\beta f(t) \tag{9.5}$$

and, thus, the instantaneous frequency is linearly related to the signal, $f(t)$.

Figure 9.3 shows the spectrum of an FM carrier where the carrier frequency is 10 MHz and the signal frequency is 1 MHz. Many upper and lower sidebands, whose amplitudes

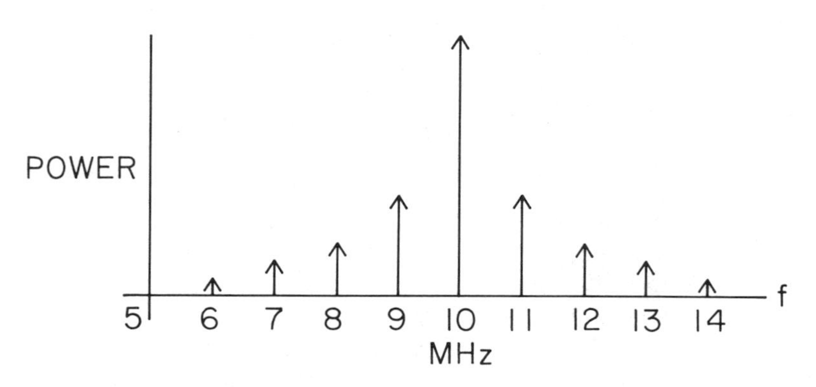

Fig. 9.3. The spectrum of a 1 MHz sinusoid when frequency modulated on a 10 MHz carrier.

depend on the modulation index, occur symmetrically about the carrier. Note that the sidebands are separated by 1 MHz, the signal frequency, and, therefore, the total bandwidth occupied by the FM signal is considerably larger than the signal bandwidth. The increased total bandwidth is the reason that FM provides high signal-to-noise ratios. As discussed in Chapter 7, noise does not correlate from one frequency to another, and, accordingly, the signal, or baseband, SNR depends on the total bandwidth used. As will be discussed later in this chapter, the video SNR depends on the system designer's choices of carrier frequency and modulation index.

There are several reasons why FM is universally used in video recorders. First, the upper and lower bandwidth ratio of video signals is much too large to be recorded directly. In the National Television Standards Committee (NTSC) system used in the United States and Japan, the video signal ranges from 30 Hz to 5 MHz, a video bandwidth ratio of almost 200,000 : 1. As was discussed in Chapter 6, a reproduce head is a spatial bandpass device. Due to the difficulties in ensuring satisfactory tape–head contact, it is not practical to make read heads with bandpass ratios greater than about 5000 : 1, which is far short of that required for video recording.

The second reason for the use of FM is that, since the video signal can be recovered by using an FM demodulator which operates by detecting zero crossings, the system can be made inherently insensitive to the deleterious effects of the output-signal amplitude instability or modulation noise that afflicts most recorders at short wavelengths. As the head–tape spacing varies, even minutely, the output amplitude changes due to the spacing loss discussed in Chapter 6. At a wavelength of 60 microinches, every microinch of spacing change causes the output amplitude to change one decibel. Moreover, as can be deduced from Equation 9.5, the effect of noise is proportional to its frequency deviation from the carrier frequency, and therefore, noise close to the carrier has little effect. The FM system rejects long wavelength noise; this is called "triangulating" the noise voltage or "parabolic weighting" the noise power. A PM system does not have this desirable property.

The third reason for the use of FM is that the video SNR can be increased. In professional video recorders the bandwidth recorded on tape is almost a factor of three greater than the video bandwidth and, due to the parabolic weighting, the video SNR increases as the square of the bandwidth.

9.3 Amplitude and Phase Equalization

In Chapter 8, the exact criteria were given for the amplitude and phase responses of a linear system required to obtain distortion free transmission. Clearly, an FM wave would pass through such a system undistorted at every point in its waveform. FM waves have, however, extremely large bandwidths; the upper and lower sidebands shown in Figure 9.3 extend, mathematically, to infinitely high positive and negative frequencies. Because recorders have an upper and a lower bandedge-limiting frequency, which is defined by the dimensions of the read head, it follows that some distortion is unavoidable in an FM recorder. The unavoidable distortion is made extremely small in practice by choosing a sufficiently low modulation index so that, for example, the lower sidebands which go through zero frequency are sufficiently low. In other words, the unavoidable distortion is made negligible (−60 dB) by making the FM wave sufficiently narrowband that it fits the recorder's bandpass.

The avoidable distortion is governed by the amplitude and phase responses of the recorder within the bandpass of the recorder. When flat amplitude is used, it is found that the video signal-to-noise ratio is too low; this is because most of the system's noises increase with frequency, or wave number.

Inspection of Figure 9.3 shows that when FM is passed through a channel that does not have a flat amplitude response, but rather slopes down in a straight-line manner as shown in Figure 9.4, the sum of the amplitudes of each pair of upper and lower sidebands remains unchanged. It turns

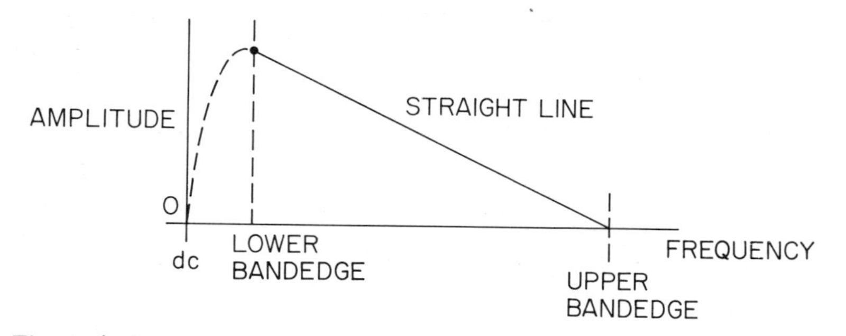

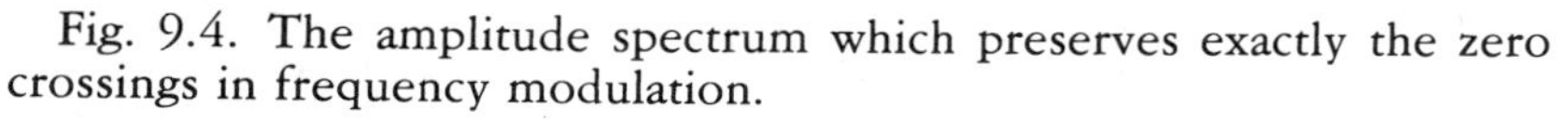
Fig. 9.4. The amplitude spectrum which preserves exactly the zero crossings in frequency modulation.

out that, when such "straight-line" equalization is used, the positions of the zero crossings of the FM wave are not changed. Accordingly, an FM demodulator, based upon such zero-crossing detection, yields an undistorted baseband, or video signal. Note that the straight-line amplitude response must occur when the output and frequency are both plotted linearly. Straight-line equalization is used in all FM video recorders because it almost matches the output spectrum of the write–read process, it minimizes the high-frequency noise, and it yields video with no avoidable distortion.

Note that straight-line amplitude equalization preserves only the positions of the zero crossing of the FM wave. The remainder of the FM waveform is distorted in a manner that is of no interest because a zero-crossing type of FM demodulator is used. An analogous situation occurs in high-density digital recording, where, as is discussed in Chapter 10, the equalization is also adjusted to give zero distortion, called intersymbol interference, at one specific sampling point in each digital bit cell. In digital recording, the particular part of the waveform that is preserved is the peak amplitude of the digital output pulses; however, since these pulses are usually detected by differentiating and finding the zero crossing, a digital system is very similar to an FM system. It is interesting to note that the FM waveforms used in video recording are virtually indistinguishable from high-density digital waveforms. It follows that no significant difference exists between the specifications of a high-density digital tape and a video tape.

The phase response sought in video recorders is, ideally, a constant group delay with a dc intercept of $n\pi$ radians. However, there are several reasons why the actual dc intercept is not critical. First, the FM signals used are relatively narrowband, and it follows that the principal effect of a 90° phase shift, as with a change from longitudinal to perpendicular recording, is to move all the zero crossings almost equally a distance corresponding to one-quarter of the wavelength of the FM carrier. Not only would this shift be imperceptible, being only a tiny fraction of a television line's length, but also, since the horizontal sync pulses are also shifted, the position of the television image does not change.

Throughout this discussion, it has been tacitly assumed that the read–write process in video recorders is linear, as defined in Chapter 8. In video recorders, of course, ac bias is not used and the write process is profoundly nonlinear. However, as is discussed at length in chapter 10, under some specific conditions, which, fortunately, usually prevail in video recorders, the magnetization written on the tape may be regarded as just the linear sum of the magnetization transitions caused by each zero crossing of the FM wave. The situation is generally called linear superposition and it is a form of pseudo- or quasi-linearity. It thus turns out that, even though the video recorder is a highly nonlinear channel, it is possible to obtain almost perfect linear system response so that the output signal is an almost undistorted replica of the input video signal.

9.4 Rotary-Head Technology

The highest frequencies in video signals are typically 5 MHz, and in order to record them at a not too-short wavelength, high head–tape relative speeds are required. In the first successful video recorder, shown in Figure 9.1, the head–tape speed used was 1500 inches per second. Current professional and consumer video recorders use 1000 and approximately 200 inches per second, respectively. These speeds are too high for reliable operation over long periods, without maintenance, of fixed-head machines.

In a rotary-head machine there are two simultaneous motions: the tape is transported in the normal manner and the heads, mounted on a drum, rotate at high speed. Figure 9.5 shows the two main types of rotary-head arrangement. In the original Ampex "quadruplex" recorders, the axis of rotation of the drum was parallel to the direction of the tape motion, and the heads, therefore, described tracks almost transversely across the tape. In later professional and all consumer machines, the drum axis is inclined so that the heads describe tracks that are more nearly parallel to the tape motion; this type of scan is called helical. In the transverse quadruplex recorders each track was, of necessity, short and only a portion (16 lines) of a video field could be recorded. Slight differences from one tape scan to the next caused visible defects in the picture. The later helical recorders overcame this problem because longer tracks could be scanned and one complete television field could be scanned on each pass.

In the long term, a most important consequence of rotary-head technology is that it separates the track density on tape from the number of heads required. In fixed-head recorders, a track density of 1000 tracks per inch over a one inch wide

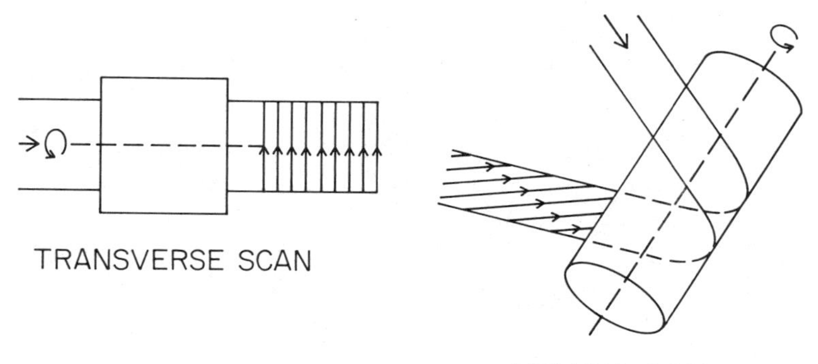

Fig. 9.5. The transverse and helical rotary-head scanning techniques.

tape mandates 1000 write–read heads. With rotary-head technology, the same track density can be achieved with a single head. If more heads are used, as in all video recorders, it is only to accommodate some other design requirement, such as, for example, simple reliable tape cassette loading, continuous recording without missing intervals, or read-after-write capability.

9.5 Machine Specifications and Parameters

In Table 9.1, the principal characteristics of two video recorders are listed. The professional machine is a helical type-

Table 9.1
Principal Parameters of Professional and Consumer Recorders

Parameter	Professional	Consumer
Video upper frequency, (MHz)	5	2
Video lower frequency, (Hz)	30	30
FM carrier frequency, (MHz)	7.9	3.9
FM sensitivity, (MHz/volt)	3	1
FM bandwidth, (MHz)	15	5.6
Video SNR, (dB)	52	42
Tape speed, (ips)	10	0.8
Head speed, (ips)	1000	220
Drum speed, (rps)	30	30
Longest wavelength, (mils)	1	10
Shortest wavelength, (microinch)	60	40
Signal optimization	max, short	max, short
Write gap length, (microinch)	30	15
Read gap length	same	same
Tape type	Co-γ-Fe_2O_3	Co-γ-Fe_2O_3
Tape width, (inches)	1	0.5
Coating thickness, (microinch)	200	200
Track width, (mils)	5.5	0.7
Guard bandwidth, (mils)	2.5	none, slant gap
Track density, (tpi)	125	1400
Video luminance	composite FM	FM
Video chrominance	composite FM	direct
Video per scan	1 field	1 field

C recorder; many thousands are used in video production and television studios throughout the world. The consumer video cassette recorder (VCR) is a Video Home System (VHS) type of which almost one hundred million examples are now being manufactured annually.

9.6 Video Signal-to-Noise Ratios

The analysis of the video, or baseband, signal-to-noise ratio of an FM video recorder is more complicated than for direct recorders because the characteristics of the FM system must be taken into account.

It turns out that the video signal-to-noise ratio is given by

$$\mathrm{SNR} = \frac{(\Delta k)^2\ \mathrm{SPS}(k_c)}{N} \tag{9.6}$$

where $\mathrm{SPS}(k_c)$ is the signal power at the carrier frequency, k_c, and N is the parabolically weighted noise power. The $\mathrm{SPS}(k)$ is as given previously in Chapter 7. The noise is weighted parabolically, that is, proportional to the square of its frequency deviation from the carrier frequency, thus,

$$N = 2 \int_{k_c - k_s}^{k_c + k_s} NPS(k) \cdot (k - k_c)^2\, dk \tag{9.7}$$

where k_s is the video signal bandwidth.

The approximate result can be given in spatial and temporal equivalent forms,

$$\mathrm{SNR} = \frac{3\pi n W (\Delta k)^2}{2 k_c k_s^3} \tag{9.8}$$

and

$$\mathrm{SNR} = \frac{3 n W\, V^2 \beta^2}{8\pi f_c f_s^3} \tag{9.9}$$

where n is the particle density in cm^{-3}, W is the track width in centimeters, V is the head–tape speed in cm/sec, Δk, k_c, k_s are the sensitivity, carrier, and video bandwidth wave numbers in cm^{-1}, and β, f_c, f_s are the sensitivity, carrier, and video bandwidth frequencies in hertz.

It is probable that the temporal form, Equation 9.9, gives the more insight. The SNR is, as might be expected, proportional to the particle density, track width, and head–tape speed squared, just as it is in direct recorders. Thus, when the FM system parameters are held fixed and the head–tape speed is doubled, the video SNR is quadrupled (+6 dB). When, with a fixed video bandwidth and carrier frequency, the modulation index is doubled, the video SNR is quadrupled (+6 dB); this is an example of the increase in SNR that is possible in FM by using a greater bandwidth. Note, finally, that by using a modulation scheme, the system designer is now able to deliver, within reasonable limits, any SNR required. Whereas in direct recorders the SNR was governed solely by tape parameters, now several FM system parameters enter the SNR expressions.

All the SNR expressions discussed above are, of course, mean-signal-power-to-mean-noise-power ratios. In the television industry, it is the habit to use a different definition for signal-to-noise ratios; this not only gives a larger number but also suits better the "peaky" nature of the video signal. The SNRs usually quoted are peak-peak signal squared to mean noise; this adds a factor of $2\sqrt{2}$ squared or 8, (+9 dB).

For a type-C professional video recorder, the calculated video SNR, obtained by substituting $n = 10^{15}$, $W = 5 \times 2.5 \times 10^{-3}$, $V = 1000 \times 2.5$, $\beta = 3 \times 10^6$, $f_c = 7.9 \times 10^6$, and $f_s = 5 \times 10^6$ in Equation 9.9, is 8×10^4, or 49 dB. Adding 9 dB to this value yields 58 dB, which is rather larger than the measured value of 52 dB. The difference is attributed principally to the effects of modulation noise.

With the VHS consumer VCR, substituting $n = 10^{15}$, $W = 0.7 \times 2.5 \times 10^{-3}$, $V = 220 \times 2.5$, $\beta = 1 \times 10^6$, $f_c = 3.9 \times 10^6$, and $f_s = 2 \times 10^6$ results in 2×10^3, or 33 dB. Adding 9 dB gives the final result, 42 dB, which is exactly the measured value.

Again it is to be noted that the relatively close numerical agreement of the calculated and measured values is, to a certain degree, good fortune. Nevertheless, the insight into the workings of an FM video recorder system that is provided by the simple expressions makes their close study worthwhile.

9.7 Slant-Track Recording

In order to achieve longer playing times for a given length of tape, all consumer VCRs use slant-track recording; this makes it possible to dispense completely with a guard band between adjacent tracks. In slant-track recording, the azimuth angles of the gaps in the heads which are used to record adjacent tracks are deliberately made significantly different.

In all recorders, the reproducing head does not follow precisely the trajectory of the writing head. This is due to a wide variety of mechanical factors. When the same head is used for both writing and reading, the tracking errors are, of course, much smaller.

In professional recorders, guard bands are placed between the adjacent tracks and they are of sufficient width to ensure that a read head never overlaps an adjacent track. In the type-C machines, the guard band is 2.5 mils (2.5×10^{-3} inches) wide, and in this case, mistracking merely reduces the signal and signal-to-noise ratio.

In the consumer machines, a statistical approach is taken

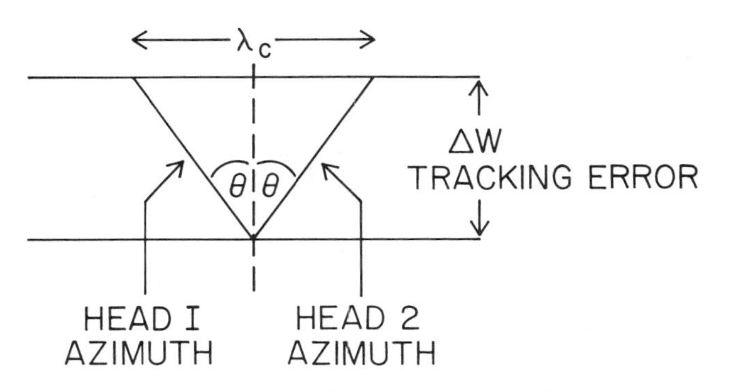

Fig. 9.6. Crossed azimuth recording of adjacent tracks.

to the problem. Suppose that the most probable off-track distance is ΔW and that viewing tests show that a particular FM wavelength, λ_c, on tape yields the most objectionable or noticeable television picture impairment. Then the gap slant, or azimuth, angles are chosen so that that off-track distance and wavelength fall into the first azimuth loss null.

As shown in Figure 9.6, this condition occurs when

$$\lambda_c = 2\,\Delta W \tan\theta \tag{9.10}$$

where $\pm\theta$ are the gap azimuth angles. Typically, in a VHS recorder, $\lambda_c = 80$ microinches, $\Delta W = 200$ microinches, and θ is about 10°.

Exercises

1. Give three reasons why FM is used in all analog video recorders.
2. A VHS recorder uses about 6 MHz bandwidth on tape for a 2 MHz bandwidth video signal. Why is this done?
3. What system transfer function is needed to preserve the zero crossings of a frequency modulated signal?
4. Why is an FM video recorder relatively immune to small amplitude variations of the read-head signal?
5. Suppose that only the head–tape speed were to be doubled in a video recorder. How would the video SNR change?
6. If an iron particle tape, with one-tenth the particle volume but the same pvc, were to be substituted for a Co-γ-Fe_2O_3 video tape how would the video SNR change?
7. What is the main reason that frequency modulation, rather than phase modulation, is used in video recorders?

8. What is the relationship between the video SNRs usually used in the television industry and the mean-signal-power-to-mean-noise-power ratio usually given in this book?
9. When a copy of a video tape is made by recording with the playback signal of the original (i.e., machine-to-machine dubbing), how great a reduction in video SNR do you expect to occur?
10. Why is the phase-response dc intercept not of great importance in video recorders?
11. If the sensitivity of an video recorder's FM system is doubled, how does the video SNR change?
12. A rotary-head recorder must use a rotary transformer to couple the rotating heads to the stationary machine. How does this affect the low-frequency response of the FM signal reproduced off tape?

Further Reading

Benson, K. B. ed. (1986). *Television Engineering Handbook.* McGraw-Hill, New York.

Felix, M. O., and Walsh, H. (1965). FM systems of exceptional bandwidth, *Proc. Inst. Elec. Eng.* **1112**, 1659–1668.

Mallinson, John C. (1976). The signal-to-noise ratio of a frequency-modulated video recorder. *E.B.U. Rev.,* **153**, 241–243.

Chapter 10

Digital Recorders

10.1 Introduction

There are two basic classes of digital recorders: those in which the input and output signals are themselves binary digital signals or data streams and those in which the input and output signals are analog material such as audio or video programs. The vast majority of digital recorders fall into the first class and are, principally, peripheral components of computing systems; these components includes millions of flexible disc drives, rigid disc drives, and tape recorders. Several thousand instrumentation recorders, which are operated as high-density digital telemetry recorders, also fall into this class. The second class of digital recorders is presently very small, consisting, worldwide, of a few hundred studio digital audio recorders and even fewer professional digital video recorders. This class of recorders is expected, however, to burgeon rapidly with the advent of the consumer rotary-head digital audio transports (R-DAT) and, eventually, consumer digital video recorders.

Digital recording is expected to become increasingly common and eventually replace analog machines. The reasons for this are precisely the same as for the increasing dominance of digital techniques in other fields. They include the low signal-to-noise ratio which is tolerable, the possibility of asynchronous operation, and the facility with which error detection and correction can be performed. It is well to remember, however, that no matter what may be the role of the highly nonlinear digital recorder or system, its function is to produce an output signal that is, as nearly as is possible, a perfect replica of the input signal.

10.2 Arctangent Transitions

In Chapter 5, the Williams–Comstock model was introduced. In this model, it is assumed that the form of magnetization transition written by a step-function-like change in the write–head current is that of an arctangent,

$$M_x(x) \propto \tan^{-1}\left(\frac{x}{f}\right) \tag{10.1}$$

whose Fourier transform is

$$M_x(k) \propto \frac{e^{-kf}}{jk} \tag{10.2}$$

where x is the longitudinal distance in centimeters, f is the arctangent parameter in centimeters, k is the wave number in cm^{-1}, and $j = \sqrt{-1}$.

The arctangent transition is shown, again, in Figure 10.1. It is clear that any one of a host of mathematical functions that range from -1 to $+1$ could have been chosen. The particular merit of the arctangent form is that the associated mathematical analysis, which is given below, yields extremely simple expressions. In order to make plain this sim-

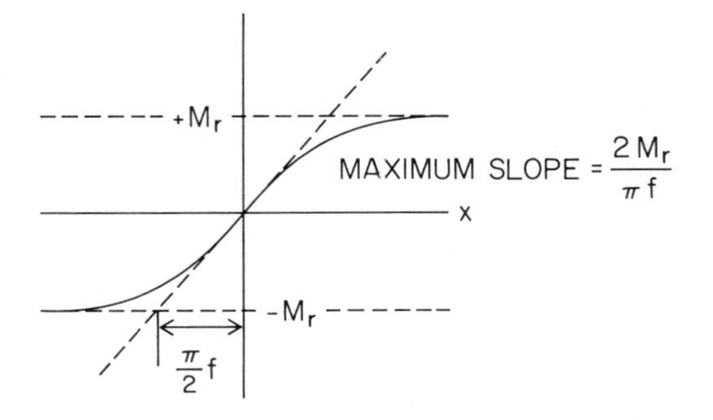

Fig. 10.1. The arctangent magnetization transition.

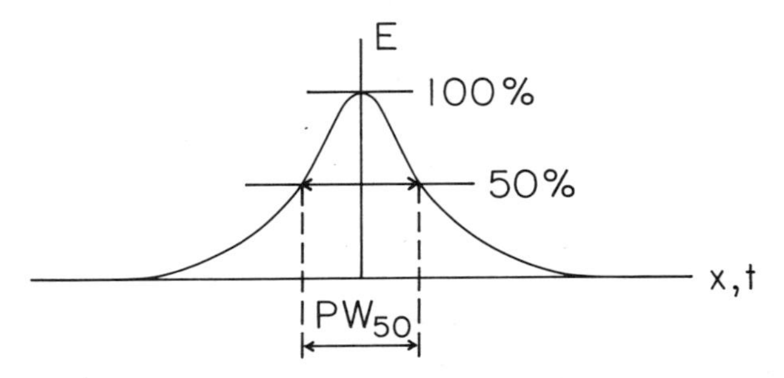

Fig. 10.2. The isolated output voltage pulse.

plicity, all constants and factors of proportionality are omitted in this chapter.

As was noted in Chapter 5, the arctangent parameter f is inversely proportional to the maximum slope. The slope is, by differentiation,

$$\frac{dM_x(x)}{dx} \propto \frac{f}{x^2 + f^2} \tag{10.3}$$

which has its maximum value at $x = 0$, thus

$$\text{maximum slope} \propto \frac{1}{f} \tag{10.4}$$

The voltage output pulse, which the arctangent transition produces in an (almost) zero gap-length read head, is

$$E(x) \propto \log_e \left(\frac{(d + \delta + f)^2 + x^2}{(d + f)^2 + x^2} \right) \tag{10.5}$$

as is shown in Figure 10.2. The width of the output pulse, measured at the one-half of maximum amplitude level, is called PW_{50} and is given by

$$PW_{50} = 2\sqrt{(d + f)(d + \delta + f)} \tag{10.6}$$

In Equations 10.5 and 10.6, d is the head–medium spacing and δ is the coating thickness of, or the depth of recording in, the medium.

For thin media, Equations 10.5 and 10.6 reduce to even simpler forms. The output pulse width becomes

$$\mathrm{PW}_{50} = 2(d + f) \tag{10.7}$$

and the output pulse takes on the so-called Lorentzian form,

$$E(x) \propto \frac{1}{1 + (2x/PW_{50})^2} \tag{10.8}$$

10.3 Linear Superposition

Under the proper conditions, it is found that a sequence of write-head current changes results in a magnetization pattern that is exactly the sum of the overlapping sequence of individual magnetization transitions. This phenomenon, where the essential nonlinearity of the write process is contained fully in the writing of each transition, is called linear superposition or quasilinearity. Note that linear superposition is not necessarily associated with arctangent transitions, even though the analysis below assumes them for the sake of simplicity.

Suppose that a sequence of equally spaced arctangent transitions is recorded, as is indicated in Figure 10.3. By linear superposition the magnetization waveform recorded is

$$M_x(x) \propto \sum_n a_n \tan^{-1}\left(\frac{x - n\Delta}{f}\right) \tag{10.9}$$

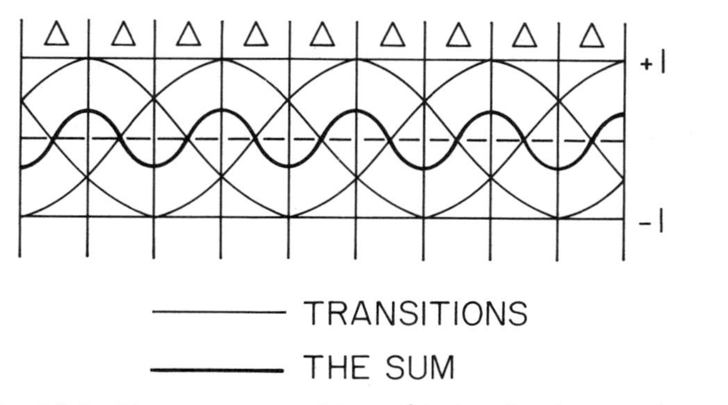

Fig. 10.3. Linear superposition of isolated voltage pulses.

where n is the transition number, a_n is ± 1 alternately, Δ is the interval between transitions, and f is the arctangent parameter.

Note that when Δ becomes smaller, corresponding to higher digital density or shorter wavelength recording, the magnetization waveform becomes progressively more nearly sinusoidal.

There are two specific conditions which must be met during the write process in order for linear superposition to hold. First, the rise time of the write-head gap field or flux must be essentially complete within the bit interval Δ. This criterion ensures that the magnetic material in each and every bit cell experiences exactly the same magnitude peak write-head field, which does not depend on the flux history of the head during previous bit cells. It follows that all transitions have the same magnitude. Note carefully that it is not sufficient to ensure that the write-head coil current rises fully within the bit cell. As discussed in Chapter 3, the losses in all magnetic materials used in heads cause the write-head gap flux to lag in phase and time behind the coil current. In order to make the write-head gap flux rise sufficiently rapidly, considerable pre-equalization of the write current is needed at high data rates. Because the write process without ac bias is nonlinear, the pre- and post-equalizers have specific and noninterchangeable roles. In unbiased recording, whether it be in video or digital recorders, the pre-equalizer has the specific task of compensating for the write-head losses.

The second condition required for linear superposition to hold is that the write-head-to-medium spacing must not be too large. When this spacing is small, the demagnetizing fields arising from the bits already written are largely imaged out by the proximity of the highly permeable write-head pole pieces. As was discussed in Chapter 5, a mirrorlike image is formed in highly permeable materials, which reduces the longitudinal components of the demagnetizing field. If the write-head–medium spacing is too large, the total field writing each bit cell is no longer of constant magnitude but varies from bit cell to bit cell depending on the demagnetizing field from previously written data. When the spacing is too large, a variety of nonlinear effects start to appear in the output voltage waveform; these effects are usually called nonlinear intersymbol interference and they lead to bit errors. It is the specific task of the recorder's post-equalizer to compensate for intersymbol interference; however, the linear filters, or post-equalizers, usually used can only correct the linear part of the intersymbol interference.

By setting the maximum permissible write-head–medium spacing equal to a fraction of a bit cell,

$$d_{\omega} \leqq 0.2\Delta \tag{10.10}$$

surprisingly good accord with widespread experience in longitudinal recording is obtained. Because designers usually choose not to deal with nonlinear intersymbol interference, the linear density achieved is inversely proportional to the write-head spacing as is shown below.

Bit density (bpi)	Bit cell (microinch)	Write spacing (microinch)
10,000	100	20
20,000	50	10
40,000	25	5
80,000	12.5	2.5

10.4 Reproduction of Square Waves

When a succession of equally spaced transitions is written, as is the case in square-wave recording, the output spectrum contains only the odd (first, third, fifth, etc.) harmonics. The amplitudes of these harmonics can be deduced easily when linear superposition holds, because the essential nonlinearity of the record channel is contained in the isolated transition itself.

When the isolated transitions have the arctangent form, the magnetization waveform is given by Equation 10.9.

Fourier analysis of this waveform yields its harmonic components,

$$M_q = \frac{4}{\pi}\frac{e^{-k_q f}}{q} \tag{10.11}$$

where M_q is the qth harmonic amplitude, q is the (odd) harmonic number, k_q is the harmonic wave number, and f is the arctangent parameter.

In Chapter 6, the linear reproduce process was discussed for the case of a purely sinusoidal tape magnetization pattern. The complex voltage transfer function of the read process for an (almost) zero-gap length read head is

$$R(k) = j(1 - e^{-k\delta})e^{-kd} \tag{10.12}$$

where $R(k)$ is the voltage transfer function, $j(= \sqrt{-1})$ and indicates the 90° phase shift, δ is the coating thickness, or depth, of recording, and d is the head–medium spacing.

The output-voltage amplitude of the fundamental ($q = 1$) of the recorded square wave is, therefore,

$$E_1 \propto \frac{4}{\pi}(1 - e^{-k_1\delta})e^{-k_1(d+f)} \tag{10.13}$$

where $k_1 = (\pi/\Delta)$ is the fundamental's wave number. At high densities, or short wavelengths, where the thickness loss term nearly equals unity, the amplitude of the fundamental becomes

$$E_1 \propto \exp(-k_1(d + f)) \tag{10.14}$$

It follows that when the fundamental's amplitude is measured in decibels and is plotted against wave number, a straight line, whose slope is proportional to $(d + f)$, results; this is shown in Figure 10.4.

The most convenient and widely used method to determine the arctangent parameter follows: measure the spectrum of a square-wave's fundamental; subtract the known or measured gap loss; measure the corrected spectral slope at high densities and subtract the known read-head–medium spacing, d. Even when the head–medium spacing is not known, the short wavelength slope of the spectrum is a very

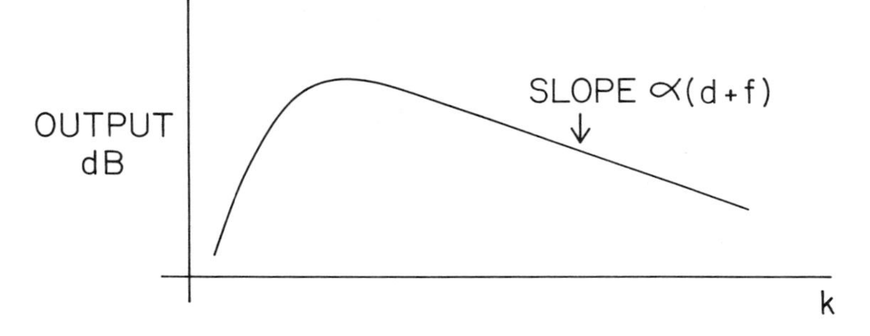

Fig. 10.4. Log-linear signal spectrum from which the total effective spacing loss may be deduced.

useful measure of the recorder's performance. The quantity $(d + f)$ may be regarded as the total, or effective, spacing.

Occasionally it is useful to consider the arctangent parameter as a hypothetical write-process spacing and to imagine that a write process spacing loss, e^{-kf}, characterizes the write process. For example, since the write-spacing loss effects the signal but not the noise, all the signal-to-noise ratio expressions discussed earlier in the book may be modified to read,

$$(\mathrm{SNR})_{\mathrm{true}} = (\mathrm{SNR})_{\mathrm{ideal}}\, e^{-2kf} \qquad (10.15)$$

where $(\mathrm{SNR})_{\mathrm{ideal}}$ is the ratio calculated with no write-spacing loss.

When the write-head current amplitude is adjusted, the optimum high density, or short wavelength, write process occurs when the total, or effective, spacing loss is as small as possible. For thick γ-Fe_2O_3 media, it is usually found that, in the optimum condition, the hypothetical write spacing is approximately equal to the actual write-head–medium spacing and the total effective spacing loss is then approximately 100 to $110d/\lambda$ dB. An approximate expression for the arctangent parameter expected in thin media was given in Chapter 5 (see Equation 5.15).

It should be noted carefully that the experimental observation of a straight line, constant slope, upper frequency spectrum in square-wave recording does not prove that precise arctangent transitions have been written or that linear superposition was valid. Any other transition shape which goes through zero like an arctangent will also yield a straight-line spectrum. A breakdown of linear superposition is nearly undetectable in square-wave recording because its principal effect is simply to change the spectral slope. Critical tests of the validity of linear superposition require the use of isolated pulses or nonperiodic sequences; a discussion of these techniques is beyond the scope of this book.

10.5 Machine Specifications and Parameters

In Table 10.1, the principal parameters of two computer peripheral digital recorders are listed. The first is a nine-track, 1600 bpi tape machine of the type most widely used for storing large data bases and exchanging data bases between computing facilities. Even though this type of machine has been superceded by more recent models, it remains among the most widely used in computer centers. The second machine is a 14-inch diameter multidisk drive representative of the present state of the art.

Table 10.1
Principal Characteristics of Two Computer Digital Recorders

Parameter	Tape Drive	Disc Drive
Data rate, (Mbs)	0.8 (parallel)	24
Digital code	Manchester	run length 2, 7
Bit density, (bpi)	1600	10,000
Head–medium speed, (ips)	60	1500
Shortest wavelength, (microinch)	600	100
Signal optimization	max. long	max. short
Write gap length, (microinch)	200	35
Read gap length, (microinch)	200	same
Head flying height, (microinch)	contact	15
Media type	γ-Fe_2O_3	γ-Fe_2O_3
Tape width/disc diameter, (inch)	0.5	14
Coating thickness, (microinch)	200	30
Track width, (mils)	80	0.7
Guard band width, (mils)	30	0.3
Number of tracks	9	800

10.6 Pulse Widths and Linear Density

The two digital recorders, whose specifications and parameters are listed above, are operated with minimal post-equalization. Whereas audio and instrumentation recorders are equalized "flat" and video recorders have "straight-line" equalization, in these computer-peripheral digital recorders, the nonlinear operation of one-zero bit detection is performed on the unequalized read-head output pulses. There are, of course, many reasons for this. The tape machine is an old design operating at a low bit density. The disc drive uses but a single electronic bit detector to serve, sequentially, as many as 20 heads. Increasing use of channel equalization is to be expected in future designs.

The output voltage pulse, due to a single arctangent transition, is given by Equation 10.5. The pulse width, measured at the 50% of peak amplitude points, is called PW_{50} and is given by Equation 10.6. In the 1600 bpi tape drive, the write current has to overwrite, or erase, the previously written data, and accordingly, it is necessary to write through the full coating thickness. The arctangent parameter is close to the coating thickness and the write head–tape spacing is negligible. Substituting $d = 0$, $f = 200$ microinch, and $\delta = 200$ microinch into Equation 10.6 yields,

$$PW_{50} = 2\sqrt{200.400} = 566 \text{ microinch} \qquad (10.16)$$

When 566 microinch wide pulses occur in sequence at the 625-microinch bit interval, which corresponds to 1600 bpi, very considerable intersymbol interference is produced. With high write current and no post-equalization, it appears this head–tape interface cannot support a substantially higher linear density because the unequalized output pulse is too broad.

In the case of the modern disc file, substitution of $d = 15$ microinch, $f = 30$ microinch, and $\delta = 30$ microinch gives,

$$PW_{50} = 2\sqrt{45.75} = 116 \text{ microinch} \qquad (10.17)$$

and it is clear that, without post-equalization, the linear density cannot greatly exceed 10,000 bpi.

Note that the PW_{50} scales with the critical dimensions, d, f, and δ, of the head–medium interface. Thus if the head–medium spacing, the arctangent parameter, and the coating thickness all are halved, the output pulse width also halves. Note also that, as can be seen in Equation 10.13, the output voltage, measured at one-half the wavelength corresponding to twice the bit density, remains unchanged. This is because the terms $k\delta$ and $k(d+f)$ are scale invariant. Similarly, in Equation 10.5, because a ratio of dimensions occurs, it is easy to see that the output pulse peak value is scale variant. These facts have dominated the design of disc drives over the last two decades. A doubling of linear density has been achievable, most easily, by simply halving the dimensions. How much further this design strategy will serve is an open question because it depends on such relatively imponderable matters as the wear, friction, and corrosion of the increasingly thin films involved.

10.7 High-Density Digital Recording

When substantial post-equalization of the read-head output is employed, considerably higher linear bit densities can be used without excessive intersymbol interference. Because post-equalization acts by emphasizing the short-wavelength parts of the signal, where the system's slot signal-to-noise ratio is lower, post-equalization always lowers the wideband signal-to-noise ratio.

Consider the case of binary digital recording on the type of instrumentation recorder discussed in Chapter 8, where the output signal has been equalized flat in amplitude and the 90° phase error has been corrected properly. The system's transfer function, shown in Figure 10.5, is idealized for this discussion so that it remains flat down to dc and is zero above the upper bandedge wavelength of 60 microinches. The impulse response of the system is shown in Figure 10.6. Note that, apart from the center peak, the output is exactly zero at intervals of one-half the upper bandedge wavelength. Thus, if a sequence of impulses were to be recorded at exactly 30

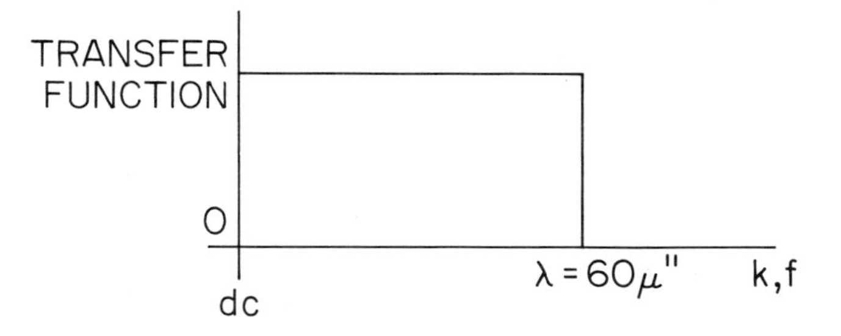

Fig. 10.5. The ideal "brick-wall" low-pass transfer function.

microinch spacings, zero intersymbol interference would occur at the 30 microinch intervals because there only one impulse in the entire sequence has a nonzero output. In digital recording, of course, step functions are recorded, but the arguments are similar and the conclusion the same. With 30 microinch bit cells, the linear density is 33,333 bpi and it corresponds to the data rate of 4 Mbs mentioned in Chapter 8.

Operation at a bit interval equal to one-half the upper bandedge wavelength can be achieved with any of the (infinite) set of transfer functions shown in Figure 10.7; all have the same symmetry about the upper bandedge. The upper bandedge is often called the Nyquist frequency, and the data rate, corresponding to two bits per cycle of the bandwidth, is termed the Nyquist rate. The Nyquist rate is the highest data rate possible for which zero intersymbol interference and,

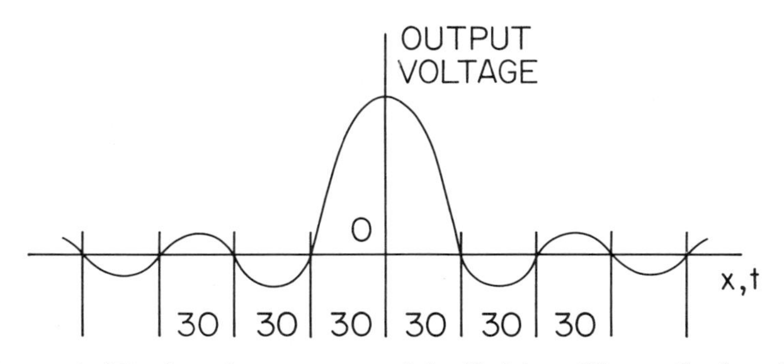

Fig. 10.6. The impulse response of the "brick-wall" transfer function.

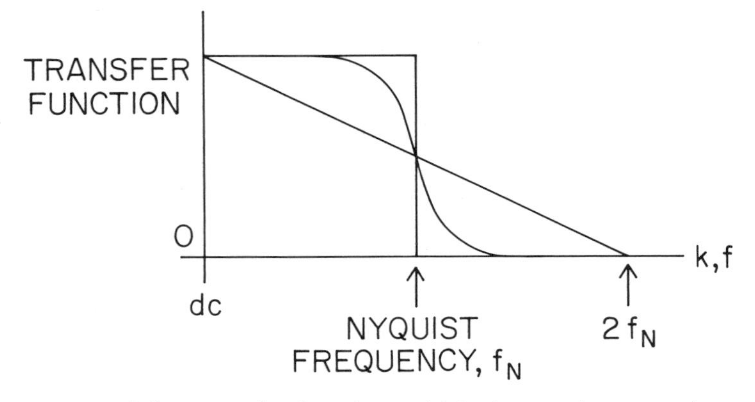

Fig. 10.7. Other transfer functions which also produce zero intersymbol interference.

consequently, independence of each bit can be achieved. The actual transfer function selected by the system designer is determined mainly by signal-to-noise ratio considerations. Typical designs include response up to about 1.5 times the Nyquist frequency. Because digital systems can tolerate some intersymbol interference, the Nyquist rate may be exceeded in principle, but in practice this rarely occurs.

10.8 Reciprocity With Very Thin Media

Consider digital recording on a very thin recording medium. As was discussed in Chapter 5, the arctangent parameter is comparable to the coating thickness. Assume that it is set equal to zero and that ideal, or perfect, step functions of magnetization are written. The output voltage, due to a step-function transition, may be calculated by the Reciprocity Integral discussed in Chapter 6, and is

$$E(x) \, \alpha \, \frac{d}{dt} \int \mathbf{h} \cdot \mathbf{M} \, dV \tag{10.18}$$

For a very thin medium of thickness δ and track width W, moving at head–medium relative speed of V, this becomes

$$E(x) \propto V\,\delta W \int \mathbf{h} \cdot \frac{d\mathbf{M}}{dx}\,dx \qquad (10.19)$$

But for perfect step functions, $d\mathbf{M}/dx$ is a delta or impulse function, and the integral is only "turned on" at the delta function. The result for longitudinal recording is

$$E(x) \propto h_x(x) \qquad (10.20)$$

The output-voltage pulse shape for the longitudinal recording of step functions on very thin media has exactly the same shape as a plot of the longitudinal component of the field above the read-head gap. For perpendicular recording, the output pulse traces a plot of the perpendicular field component.

This particularly elegant application of the Reciprocity Integral provides a result of great convenience and simplicity. The isolated pulse shapes obtained with very thin media can be regarded as displays of the read-head fringing field components, no matter what may be the construction or gap length of the head. Alternatively, when the head field shape is considered to be known, the output pulse shapes provide clues as to the direction of magnetization in the very thin medium. As was noted, however, in Chapter 6, a unique determination of the magnetization is never possible by this means.

10.9 Bit Error Rates and SNRs

Two of the main reasons for the increasing popularity of digital systems are that relatively low channel signal-to-noise ratios can be tolerated and that error detection and correction can be implemented easily. These properties arise because, with binary digital signals, the bit detector has only to decide whether a pulse is positive or negative and because, when it is believed that a given bit is in error, a simple polarity inversion corrects it.

The signal amplitude and channel signal-to-noise ratio are only important to the extent that they effect the proper performance of the bit detector. Errors in the bit detector are called "raw" bit errors. After error detection and correction (EDAC), any bit errors that remain are termed uncorrected. In computer-peripheral digital recorders, the raw bit error rates (BERs) are about 10^{-9} and are usually corrected to about 10^{-13}. Thus a disc drive running at 24 Mbs, with a 10% duty cycle, makes but one or two uncorrected bit errors per year. In a digital recorder handling analog signals, much greater BERs are tolerable; typically the raw rate is 10^{-6} and, after EDAC, the uncorrected rate is about 10^{-10}. It follows that a digital audio recorder with data rate of approximately 2 Mbs makes an uncorrected error about every two hours.

In a computer-peripheral digital recorder, the input signal is a digital data stream and uncorrected bit errors are, of course, the only faults possible in the system. With digitized analog recorders, interest centers on the effective signal-to-noise ratio of the final analog signal. In these systems, the analog signal is first passed through an analog-to-digital converter, which takes amplitude samples of the waveform at a sufficiently high rate. Ideally, a sampling rate equal to only two times the analog signal's upper bandedge is required by the Nyquist Sampling Theorem; but in practice, higher rates are used. Commonly used rates for 20 kHz audio are in the range of 44–50 thousand samples per second. The amplitude of the samples is then computed as an N-bit word, which is recorded. Upon reproduction, a complementary device, called a digital-to-analog converter, changes the digital words back to amplitude samples and these are then low-pass filtered to restore the original analog waveform.

The effective signal-to-noise ratio in such a system depends only on the number of bits used in digitizing each sample. The noise, called quantizing noise, arises because of the (small) errors made in matching the amplitudes of the samples to the 2^N levels available with an N-bit word. When the amplitude of a sample falls above the midpoint of the interval (2^{-N}) between these levels, it is assigned the upper level; when the amplitude falls below the midpoint, it is assigned the lower level. For random amplitude samples, the mean-square quantizing error is one-twelfth the interval. It follows that, since the mean power of a sinusoidal signal is one-half its peak power, the mean signal power-to-mean quantizing noise power ratio is

$$(\mathrm{SNR})_{\mathrm{quant}} = \tfrac{1}{2} \cdot 2^{2N} \cdot 12 = (6N + 7.8)\ \mathrm{dB} \qquad (10.21)$$

If, for example, $N = 8$ as it is in digital video recorders, $(\mathrm{SNR})_{\mathrm{quant}} = 55.8$ dB; in digital audio recorders, 16-bit resolution is used and $(\mathrm{SNR})_{\mathrm{quant}} = 103.8$ dB.

It is interesting to note that Equation 10.21 has no terms in it which pertain to the write–read process. In direct recorders, discussed in Chapter 8, the wideband signal-to-noise ratio depends entirely on the physical parameters of the read–write process. In video recorders, treated in Chapter 9, an analog modulation scheme, frequency modulation, is used and the signal-to-noise expressions contain both write–read parameters and FM system parameters. When, however, a digital modulation scheme, or as it is often called, pulse code modulation (PCM), is used, the signal-to-noise ratio depends only on the parameters of the digitizer. In PCM, the system designer is freed from the analog constraints of the system. Provided that the recorder delivers a low enough bit error rate, the designer can provide whatever SNR is required.

Exercises

1. What is the principal function of the pre-equalizer in high rate digital recording?

2. What is the PW_{50} of the unequalized, isolated output pulse resulting when an arctangent transition is recorded on a tape of thickness δ and reproduced with an (almost) zero gap-length head at a head–medium spacing of d?

3. When recording square waves, what is the slope of the short-wavelength spectrum on a log-linear plot after it has been corrected for the usual gap-loss terms?

4. What is the Fourier transform of a perfect step function?
5. What is the linear density which corresponds to the Nyquist rate in a recorder which is equalized properly about the Nyquist wavelength of 40 microinches?
6. Why cannot the roles of the pre- and post-equalizers be exchanged in an unbiased recorder?
7. What happens if the Nyquist rate is exceeded in digital recording?
8. The voltage-output pulse, $E(x)$, arising from arctangent transitions of longitudinal magnetization on a thin medium, has the so-called Lorentzian shape. What is the mathematical form of this shape?
9. If a digital recorder is operated at 100,000 flux reversals per inch, what is the write head–medium spacing likely to be?
10. For the digital recorder in question 9, what is the total effective spacing, as measured from the high-frequency spectral slope, likely to be?
11. Sketch some magnetization transition functions that would give constant slope, upper frequency spectra, and show how they relate to an arctangent function.
12. Why is binary digital technology preferred to ternary or higher digital methods?

Further Reading

Coleman, C., *et al.* (1984). High data rate magnetic recording in a single channel. *Int. Conf. Video and Data Recording, IERE Proc.* **59,** 151–157.

Gibby R. A., and Smith, J. W. (1965). Some extensions of Nyquist's telegraph transmission theory. *Bell System. Tech. J.* **44,** 1487–1510.

Mallinson, John C. (1974). On extremely high density recording. *IEEE Trans. Mag.* **10,** 368–373.

Mallinson, John C. (1975). A unified view of high density digital recording theory. *IEEE Trans. Mag.* **11,** 1166–1169.

Mallinson, John C., and Steele, C. W. (1969). Theory of linear superposition in tape recording. *IEEE Trans. Mag.* **5**, 886–890.
Mee, C. D., and Daniel, E., eds. (1987). *Magnetic Recording,* Vol. 2: *Applications.* McGraw-Hill, New York.
Middleton, Barry K., and Wisley, Paul L. (1978). Pulse super-position and high density recording. *IEEE Trans. Mag.* **14**, 1043–1050.
Wood, Roger, *et al.* (1984). An experimental eight-inch disc drive with one hundred megabytes per surface. *IEEE Trans. Mag.* **20**, 698–702.

Appendix

Magnetic Properties in Cgs–Emu and SI Units

Quantity	Symbol	Cgs–emu	Conversion factor, C	SI
Magnetic flux density	$\mathbf{B}$	gauss	10^{-4}	tesla
Magnetic flux	ϕ	maxwell	10^{-8}	weber
Magnetomotive force	mmf	gilbert	$10/4\pi$	ampere
Magnetic field	$\mathbf{H}$	oersted	$10^3/4\pi$	A/m
Magnetization	$\mathbf{M}$	emu/cm^3	10^3	A/m
Magnetization	$4\pi\mathbf{M}$	gauss	$10^3/4\pi$	A/m
Specific magnetization	σ	emu/g	1	A · m^2/kg
Magnetic moment	$\boldsymbol{\mu}$	emu	10^{-3}	A · m^2
Susceptibility	χ	dimensionless	4π	dimensionless
Permeability	μ	dimensionless	$4\pi \times 10^{-7}$	H/m
Demagnetization factor	N	dimensionless	$1/4\pi$	dimensionless

To convert Cgs–emu to SI multiply by the conversion factor, C.

Answers to the Exercises

Chapter 1

1. $H = 4\pi$ Oe
2. $\mu_B = \dfrac{eh}{4\pi m} = \dfrac{e\hbar}{2m} = 0.9 \times 10^{-20}$ emu
3. 26, $4\mu_B$ (free space), $2.2\mu_B$ (solid state)
4. $\dfrac{4\pi}{3}$
5. $\dfrac{2}{3} 4\pi M = \dfrac{8\pi M}{3}$
6. **M** is the volume average of the magnet moments, that is,

$$\mathbf{M} = \frac{1}{V} \sum_{i=1}^{N} \mu i$$

7. $B = 4\pi M \cos\theta$, in the plane of the plane.
8. Zero
9. They are orthogonal at all points in space.
10. Amperian currents give the **B** field, while real currents give the **H** field.
11. Zero
12. Magnetic poles and real currents

Chapter 2

1. $H_c \approx M_s$, independent of size
2. Switching energy equals 50 times thermal energy
3. about 2×10^{-17} cubic centimeters
4. 1000 Oe, normal to the plate
5. ${}_bH_c < {}_mH_c < {}_rH_c$
6. Irreversible phenomena

7. That it becomes single-valued
8. $\mu = 1 + 4\pi\chi$
9. Because the demagnetizing field is zero at ${}_{m}H_{c}$
10. $\rho = -\nabla \cdot \mathbf{M}$
11. To reduce the self-, or demagnetizing, energy
12. $H_c \approx M_s = 1700$ Oe

Chapter 3

1. $(4\pi 370)(0.4)(0.8) = 1487$ G
2. $\tau = 10^{-9} \exp 30 = 1.06 \times 10^4$ seconds
3. Zero everywhere
4. There is no fundamental difference!
5. The oxidation state of the iron atoms

$$\gamma\text{-}Fe_2O_3 = Fe_2^{+++}O_3^{--}$$

$$Fe_3O_4 = (Fe^{++}O^{--})(Fe_2^{+++}O_3^{--})$$

6. This is no chemical difference.
7. $M = \dfrac{1000}{V}$ G
8. Voids, dislocations, grain boundaries, mechanical strains, and their irregular shape
9. When one of the cartesian coordinate axes coincides with the axis of revolution
10. 4π
11. $3d$
12. Because the two sublattice magnetizations are equal and opposite

Chapter 4

1. The write-head efficiency is that fraction of the coil mmf which appears across the gap.
2. $R = \dfrac{l}{A\mu}$
3. Voltage or electromotive force (emf)
4. $\text{Eff} = \dfrac{R_{gap}}{R_{gap} + R_{core}}$
5. False
6. The efficiency of low-efficiency heads is increased, while that of high-efficiency heads is little changed.

7. The pole tip corners adjacent to the top of the gap
8. 12% of the gap length
9. $H_x = \frac{H_o}{\pi} \tan^{-1}\left(\frac{yg}{x^2 + y^2 - g^2/4}\right)$

   $H_y = \frac{H_o}{2\pi} \log_e\left(\frac{(x - g/2)^2 + y^2}{(x + g/2)^2 + y^2}\right)$
10. $H_o = \text{Eff} \cdot \frac{0.4\pi NI}{g}$
11. $2NI \cdot \text{Eff}$
12. **B** everywhere

Chapter 5

1. All of them are!
2. Because the remanent flux written by a sinusoidal current is also sinusoidal
3. Position displacements, or phase errors, at different depths in the coating
4. $l = (\Delta H)\left(\frac{dH}{dx}\right)^{-1}$
5. $L[ax_1(t) + bx_2(t)] = aL[x_1(t)] + bL[x_2(t)]$
6. Because the range of switching fields, ΔH, is usually proportional to the coercivity
7. 375 Oe
8. 4350 Oe
9. Inversely
10. $f = 2\sqrt{\frac{M_r\delta}{{}_rH_c}\left(d + \frac{\delta}{2}\right)}$
11. A good vector, or two-dimensional, model of hysteresis
12. $\frac{dM_x}{dH} = \chi\left[\frac{dH_h}{dx} + \frac{dH_d}{dx}\right]$

Chapter 6

1. $20 \log_{10}[\exp(-kd)] = \frac{20 \log_e[\exp(-kd)]}{\log_e 10}$

   $\frac{20(-2\pi d/\lambda)}{2.303} = \frac{-54 \cdot 6d}{\lambda} \text{ dB}$
2. In phase
3. 90° phase correction by either differentiation or integration

4. $\text{Eff} = \dfrac{R_{\text{gap}}}{R_{\text{gap}} + R_{\text{core}}}$
5. Only that region whose magnetic poles straddle the gap produces useful flux.
6. All as N^2
7. (a) $\pm 180°$, (b) $\pm 180°$, and (c) $+90°$
8. $-\dfrac{1}{2} \cdot \dfrac{4\pi M}{3} \cdot M \cdot V = \dfrac{2\pi M^2 V}{3}$
9. Long wavelength, overall dimensions, $\sqrt{LD}$; short wavelength, gap length, g
10. Because perfectly rectangular and straight gap edges cannot be made
11. Because the internal, or self-demagnetizing, field is not uniform
12. (a) the flux entering the head

Chapter 7

1. The $(\text{SNR})_w$ is halved (-3 dB change).
2. The $(\text{SNR})_w$ is reduced by 6 dB, because the shortest wavelength is halved.
3. The desire to keep the recording medium noise power substantially higher than the electronics or other system noise powers
4. Length $= \frac{1}{2}$ shortest wavelength; depth $= \frac{1}{3}$ shortest wavelength; and width = track width
5. 2000
6. Because noises are uncorrelated, or incoherent, in time
7. An increase by a factor of 2^3(+9 dB)
8. Because integration over the noise power spectrum is addition of noise powers
9. It does not affect the SNR!
10. Noise power $\alpha\ R\ \alpha\ L^*\ \alpha\ N^2 A\ \mu^*/l$ which is proportional to the size.
11. The variance or standard deviation squared
12. No, because it would affect the signal and noise powers equally.

Chapter 8

1. Equal amplitudes at all significant frequencies; termed "flat" amplitude
2. Linear phase, or constant group delay, with an integer number of radians zero frequency intercept

3. Consider N-bit words: each has N bits of information, there are only 2^N different words, the probability of any word's occurrence is 2^{-N} and, finally, $-\log_2(2^{-N}) = N$
4. By not having "flat amplitude" and "linear phase" responses
5. Distortion only
6. To reduce the differences between the high and low ac bias output spectra
7. 1000 times wider, or 100 inches!
8. Because the maximum signals are limited to a small fraction (about one-third) of the tape's remanence
9. The reciprocal of the recorder's signal power spectrum
10. The greater the post-equalization, the lower the $(SNR)_w$.
11. Pre-equalization
12. 5×10^6 bits/second

Chapter 9

1. To accommodate the 200,000 : 1 bandwidth ratio of television; to provide immunity to small reductions in signal amplitude; and to increase the video SNR
2. To provide a higher video SNR
3. When amplitude and frequency are on linear scales, the amplitude transfer function must be straight line.
4. Because small amplitude variations do not significantly change the zero-crossing positions of the FM signal
5. A factor of four increase, +6 dB
6. A factor of ten increase, +10 dB
7. In FM, $\omega_i = \omega_c + 2\pi\beta f(t)$, and so the noise voltage is trangulated by the frequency deviation from the carrier. This parabolic noise power weighing minimizes low-frequency noises.
8. Industry uses peak-to-peak signal power and not mean signal power; accordingly, a correction factor of $2\sqrt{2}$ squared, +9 dB is to be added to the SNR derived in this book
9. It will be halved, −3 dB
10. The FM signal is sufficiently narrowband that a phase change of 90° merely shifts all zero crossings equally by one-quarter of the carrier wavelength; not only is this imperceptible, but also the horizontal sync pulses automatically correct the television receiver display.
11. A factor of four increase, +6 dB
12. Transformers cannot couple dc signals, thus there are

now three reasons for zero dc response: zero head flux, zero rate of change of head flux, and zero transformer coupling.

Chapter 10

1. To get the write-head gap flux up to its full value within the minimum time between transitions
2. $PW_{50} = 2\sqrt{(d + f)(d + \delta + f)}$
3. The slope is proportional to the total effective spacing.
4. $[jk]^{-1}$
5. 50,000 flux reversals/inch
6. Because the pre-equalizer has a specific task to perform (see answer 1 above)
7. Intersymbol interference is inevitable; note that this does not preclude error-free detection of digital signals.
8. $E(x) = \dfrac{1}{1 + (2x/PW_{50})^2}$
9. Equal to or less than 2 microinches
10. Equal to or less than 4 microinches
11. Any function, going from -1 to $+1$, that goes through zero as does an arctangent function
12. Binary digital requires the lowest possible SNR and the EDAC is the simplest possible

Index